WHAT'S SO CONT
GENETICALLY MO

FOOD CONTROVERSIES

SERIES EDITOR: ANDREW F. SMITH

Everybody eats. Yet few understand the importance of food in our lives and the decisions we make each time we eat. The Food Controversies series probes problems created by the industrial food system and examines proposed alternatives.

Already published:

Fast Food: The Good, the Bad and the Hungry Andrew F. Smith
What's So Controversial about Genetically Modified Food? John T. Lang

WHAT'S SO CONTROVERSIAL ABOUT GENETICALLY MODIFIED FOOD?

JOHN T. LANG

REAKTION BOOKS

Published by Reaktion Books Ltd
Unit 32, Waterside
44–48 Wharf Road
London N1 7UX, UK
www.reaktionbooks.co.uk

First published 2016
Copyright © John T. Lang 2016

Printed and bound in Great Britain
by Bell & Bain, Glasgow

A catalogue record for this book is available from the British Library

ISBN 978 1 78023 668 1

CONTENTS

INTRODUCTION
GENETICALLY MODIFIED FOOD: REMAKING THE GLOBAL FOOD SYSTEM

What is food to one man may be fierce poison to others.

Lucretius (99–55 BC)

If we are going to live so intimately with these chemicals, eating and drinking them, taking them into the very marrow of our bones – we had better know something about their nature and their power.

Rachel Carson, *Silent Spring* (1962)

In 1825 the French gastronome Jean Anthelme Brillat-Savarin famously remarked, 'tell me what you eat, and I will tell you what you are.' More recently, Alfred E. Neuman, the fictional cartoon mascot from the cover of *Mad* magazine, remarked: 'we are living in a world today where lemonade is made from artificial flavors and furniture polish is made from real lemons.' If what we are eating is not natural, then what, in fact, are we eating, and what does it say about us?

In the twentieth century, three pivotal scientific discoveries and one milestone conference initiated the modern biotechnology industry. In 1953 James Watson and Francis Crick published their paper determining the structure of

DNA, the genetic instructions found in all known living organisms, and in it they presented the groundbreaking double-helix model illustrating how genetic material passes from one generation to the next. Then, in 1973, Stanley Cohen and Herbert Boyer pioneered a recombinant DNA technique that transferred DNA from one bacterium into the DNA of another. This meant that scientists could bring together genetic material from multiple sources to create DNA sequences that would not otherwise be found in biological organisms. Next, in February 1975, the future Nobel Prize winner Paul Berg convened the Asilomar International Congress on Recombinant DNA Molecules. The Congress's purpose was to discuss the potential ethical and scientific hazards of biotechnology as well as appropriate strategies for its regulation. The participants outlined a set of standards for genetic research based on the degree of perceived risk, trying to balance the freedom of scientific inquiry with the new technology's real-world implications.

Finally, in 1983, thirty years after Watson and Crick's discovery, four research groups working independently published news that they had successfully placed functional genes into plant cells. This profound shift in our scientific abilities gave rise to an impressively rapid series of scientific developments and more sophisticated recombinant DNA techniques. The last thirty years have been an exceptional era for both science and the public discussion of science policy. The field of biotechnology has undergone many phases of transformation, including the birth of Dolly the cloned sheep in 1996 and the initial draft of the human genome first published by Lander et al. in *Nature* in 2001. Cross-species cloning, weaponized anthrax, stem-cell research, genetic profiling and the genetic manipulation of seeds that produce the key crops in our food system are but a few of the ways in which biotechnology advancements influence our lives.

In some ways, this progress is the logical extension of the early experiments of Gregor Mendel in the 1850s, in which he demonstrated that cross-breeding animals and plants could favour certain desirable traits. For centuries, farmers have used selective breeding to modify the genes of organisms in order to improve crops. Now, this conventional method of hybridization often involves the use of molecular markers, which are gene or DNA sequences associated with a particular gene or trait complete with a known location on a chromosome. These markers are used to track plants' genetic make-up during the variety development process. Each successive hybridized generation can have tens of thousands of new gene variants, and maybe even new genes that are different from their parents. Breeders try to mix favourable traits from both parents, but generally the progeny contain a mix, inheriting both the good and the bad traits. Breeders perform this mixing over a number of breeding cycles to build on the positive traits and minimize the negative ones. As a result, conventional breeding tries to breed out unwanted genes while breeding in those that are considered desirable.

Conventional plant breeding, marker-assisted or otherwise, winds up swapping many genes at once. As a result, researchers often create thousands of plants, generation after generation, that they inevitably end up destroying before finding one that results in a useful or noteworthy change.

As a short cut to this time-intensive process, researchers use mutagenesis. Breeders who use this technique expose seeds to chemicals or radiation that cause bits of DNA to copy incorrectly, in the hope that a useful transformation will occur. These mutations may or may not have naturally occurred over time without this intervention, but mutation breeding accelerates the process. This method has produced commercial varieties of rice, wheat and barley, as well as the prized Rio Red grapefruit.[1] Even though mutagenesis alters the genetic properties of an organism and has been used for many more decades than genetic modification, it has not been subjected to the same level of popular scrutiny.

That being said, these techniques are not what people are referring to when they colloquially say that something is 'genetically modified'. In some ways, this shows the blurred line between biotechnological and non-biotechnological breeding. When people refer to genetically modified crops, they are talking about a transgenic process where a scientist purposefully and directly manipulates a gene or DNA sequence using recombinant DNA technology. Scientists can transplant genes between organisms that are unable to breed naturally, such as crossing bacteria with rapeseed (canola). This is called transgenesis, which is scientifically distinct from cisgenesis, where seed scientists directly swap genes between two crops – two types of wheat, for example – that could otherwise naturally breed. Regardless of the genetic technique, whether using mutagenesis, conventional hybrid-ization, cisgenesis or transgenesis, seed scientists are generally pursuing the same goal – to introduce new genetic character-istics to an organism in order to increase its usefulness.

There are several possible methods of genetic modification, but scientists most commonly use two methods to
genetically engineer plants. In the biolistic method, also
known as the 'gene gun' method, scientists introduce DNA
into target tissues by accelerating a DNA–particle complex
(made up of tungsten or gold particles coated with the
desired DNA) in a partial vacuum and placing the target
tissue within the acceleration path. In other words, the gene
gun shoots a metal pellet coated with DNA into a plant tissue
target. As a result of the collision, a few genes end up incorporated into the nucleus of the plant tissue. Scientists can
also use a naturally occurring soil microbe, *Agrobacterium
tumefaciens* or *A. rhizogenes*, to introduce foreign genes into
plants. Genetic engineers use a plasmid – a piece of DNA –
and then take advantage of *Agrobacterium*'s natural abilities
to transform plant cells. This *Agrobacterium*-mediated
transfer of genetic material allows scientists to introduce
new DNA into plant cells.[2]

Scientists have been exploring and bundling DNA
sequences as part of scientific research for more than thirty
years, and it is a relatively precise method; however, there
are parts of the process that are still not fully understood or
controlled. For example, scientists cannot control where the
Agrobacterium inserts its DNA bundle. There is a chance that
this bundle can fracture, or that unwanted genes could also
be triggered. In reality, however, conventional plant breeding
is also filled with considerable uncertainty, and these same
uncontrolled events can even occur during normal breeding.
Moreover, conventional inter-variety variation and environmental effects have had more impact on a plant's genetic

expression than the scientifically introduced changes.[3] Optimists might conclude that transgenic science is no more fraught with peril than traditional breeding programmes. Pessimists, on the other hand, would be quick to point out that the absence of readily observable adverse effects does not mean that the potential for such consequences can be completely ruled out.

These transgenic processes are different from conventional breeding methods in a couple of significant ways. Transgenic techniques make it easier for scientists to isolate genes and to introduce new traits without simultaneously introducing many other potentially undesirable traits. Moreover, in using these techniques, scientists can cross biological boundaries that could not be crossed by traditional breeding, for example, transferring traits from bacteria or animals into plants. In light of these differences between conventional breeding and transgenic methods, the advancements of the latter opened up scientific possibilities that had yet to be realized for genetically altering organisms.

For purposes of genetically modified (GM) food, scientists might try to increase the yield of a crop, introduce a novel characteristic or produce a new protein or enzyme. Scientists can also give crops increased resistance to environmental and biological stresses such as heat, drought, soil nutrient deficiencies, insects and diseases. To date, however, the principal agricultural biotechnology products that have been marketed are crops genetically modified to tolerate particular herbicides and resist specific pests. The best-known example of insect resistance is the use of *Bacillus thuringiensis* (Bt) genes in maize (corn) and other crops. *B. thuringiensis* is a naturally occurring bacterium which produces crystal proteins that are lethal to insect larvae. These proteins have been transferred into corn, which has enabled the corn itself to produce its own pesticides against insects such as the European corn borer, thereby providing protection throughout the entire plant.

In addition to worrying about insect pests, farmers need to make sure that their crops are protected from weeds. Normally, farmers spray herbicides before the plants and weeds start to emerge from the soil, but plant geneticists have now developed herbicide-tolerant crops to survive certain herbicides that previously would have destroyed the crop along with the targeted weeds. Because the crop tolerates the herbicide, farmers can use that particular herbicide to destroy already established weeds. The most common herbicide-tolerant crops are resistant to glyphosate, a herbicide that is effective on many species of grasses, broadleaf weeds and sedges.

During the 1980s and '90s, biotechnology became a booming industry as a result of these scientific developments, moving from the laboratory to farms. After years of public debate, a series of protests and several legal setbacks, Advanced Genetic Sciences (AGS) conducted, in 1987, the first authorized field test of a genetically modified bacterium. Frostban, also known as ice-minus, successfully reduced frost formation but was never used commercially, for the simple reason that it was no more effective, economical or commercially viable than conventional techniques. But the test itself was revolutionary. As the first release of a genetically modified organism (GMO) into the environment, the process pushed regulatory policy and spurred the protests of anti-GMO activists.

At the end of the 1980s, the first GM foods successfully cleared the U.S. regulatory process to become a commercial reality. The first product approved by the U.S. Food and Drug Administration (FDA) was chymosin, an enzyme used in the production of firm cheeses. Estimates suggest that 70 per cent or more of cheese made in the United States is now produced using genetically engineered chymosin. Recombinant bovine somatotropin (rBST), a growth hormone given to cows to increase milk production, followed chymosin in 1993. Nevertheless, farmers' use of rBST in dairy production has been, and continues to be, modest. In 1994 Calgene's 'Flavr Savr' tomato was marketed directly to consumers with the benefits of genetic modification. Initially bearing a voluntary GM label, the Flavr Savr tomato eventually failed commercially due to a lack of sales and production difficulty. Although there was some initial excitement, and little public concern, the product never sold well and by 1997 had been taken off the market.

Following these early examples of commercialized genetic modification, the introduction of commodity crops containing genetically modified traits made GM food widely available. In 1996, the first year of widespread commercialization, farmers in six countries planted more than 1.5 million hectares of crops containing GM traits. By 2014, farmers in 28 countries planted more than 180 million hectares (roughly equivalent to the combined land mass of France, Japan, South Korea, Spain and the United Kingdom) of crops containing GM traits.[4] Worldwide, almost 80 per cent of all soybeans, a little less than three-quarters of all cotton, under one-third of all maize (corn) and almost a quarter of all canola grown are genetically modified.[5] The five largest producers of crops with GM traits, accounting for 90 per cent of the worldwide acreage, are the United States, Brazil, Argentina, India and Canada. Including the 28 countries that planted crops with GM traits, 65 countries in total have approved GM crops for food use, feed use, environmental release or planting.[6]

In 2005 the Pew Initiative on Food and Biotechnology estimated that three-quarters of all processed foods in the U.S. contained a genetically modified ingredient. Now, according to the United States Department of Agriculture, genetically modified varieties make up roughly 90 per cent of all soybeans, more than 75 per cent of cotton and more than 80 per cent of corn planted in the United States. Food manufacturers use these crops and their derivatives – such as high-fructose corn syrup, cornstarch and soy lecithin, as well as canola, soybean and cottonseed oils – as ingredients in a vast array of processed foods. Moreover, GM sugar beet was first approved for planting in 2005, yet today more than one-third of sugar used in the United States comes from a genetically modified source. Because the genetic modification of key crops used in the majority of processed foods has greatly increased since Pew's report, the three-quarters estimate is likely conservative. To put this in perspective, consider the fact that the average supermarket stocks 30,000 to 40,000 food and beverage products;[7] assuming that three-quarters of processed food contains a genetically modified ingredient, the average U.S. supermarket stocks somewhere between 22,500 and 30,000 products that contain GM ingredients.

Despite publicized debate and the widespread presence of genetically modified crops in the food supply, the idea of genetic modification has yet to fully infiltrate the public imagination. When in 2003 Rutgers University researchers asked people to recall the first image or thought that came to mind when thinking of the terms 'biotechnology', 'genetic modification' or 'genetic engineering', almost one in three could not produce a single thought or image related to these words.[8] While the introduction of GM food into the food system has brought about reactions from artists, activists, ethicists and cultural theorists as well as from scientists, regulators and industry representatives, many people lack even a clear impressionistic image or feeling associated with these advances in biotechnology, let alone a sophisticated understanding or knowledge of the processes involved. Perhaps this unfamiliarity is why GM food is so controversial.

At one level, the controversies surrounding GM food are about whether genetic modification poses a risk to human health and the environment. This perspective is at the heart of traditional approaches to regulating GM food production, and these concerns over risks seemingly hinder the growth of the biotechnology industry. However, biotechnology producers and users prefer to frame the debate in this way because it appeases the public by allowing critique or reform of GM food instead of opposition to the system as a whole.[9] Critics have raised several environmental and human health safety concerns. However, a high degree of uncertainty surrounds these concerns. Although there has been no serious environmental damage and no conclusive evidence of harm to human health due to GM food, firm conclusions about the safety of GM food would be hard to come by given that long-term threats to biological diversity and ecosystems are difficult to assess. For this reason, among others, many environmentalists call for the precautionary regulation of GM food to limit long-term, potentially irreversible harm. The precautionary principle of regulation, where agribusiness must prove that GM food causes no harm, is meant to deal with incomplete scientific knowledge in risky domains. In essence it requires inaction when faced with scientific uncertainty. Under this standard, nations would assume that genetically modified crops would cause harm until this could be proven otherwise – that is, until there is evidence showing that GM food does not damage our health or the environment. Often described as 'sound science' versus the 'precautionary principle', in the context of GM food, the North American approach to risk and uncertainty, allowing products on the market until they are demonstrated to be harmful, is a more risk-tolerant strategy than the policy adopted in Europe.[10]

While agribusiness has portrayed genetic modification as scientific progress, anti-GM activists have attempted to reframe it as perverted science. Paul Lewis, a professor of English at Boston College, is often credited with inventing the neologism 'Frankenfood' to describe GM food, which conjures a blend of Mary Shelley's novel *Frankenstein* with food.[11] In addition to creative wordplay, the widespread use of potent visual images and symbolic actions by anti-GM campaigners has generated considerable media attention since the 1990s, helping to increase the public's awareness of, and raise their ire about, genetically modified foods entering the food supply.[12]

At a deeper level, the discussions surrounding GM food touch on larger issues relating to social and political power, cultural values, corporate responsibility and intellectual property. Who should regulate GM foods, and how should they be regulated? A myriad of competing interests make these questions difficult to answer. Another series of questions related to this concerns the types of risks and benefits that stakeholders consider when assessing and managing GM-food-related hazards. How is environmental harm measured? What are the long-term human health implications of this form of agriculture? Are economic considerations paramount or are cultural values more important? Who accrues most of the benefits? Would the risks disproportionately harm a particular group? Stakeholders' final considerations can have implications not only for themselves, but for ordinary people as well. Given the globalization of GM food, food production and agricultural trade, no individual or organization can hope to address these complex questions in isolation.

In short, several reasons underlie consumers' reactions, of which there is a wide-ranging spectrum, to GM food: many people remain unaware of how much GM food exists in their local shops, some experience anxiety and fear, some tenuously accept its presence in the food supply and some are explicitly opposed to it. Though some experts and partisans may accuse opponents of harbouring irrational fears about GM food, the public is quite rational in its own way. Most people do not try to evaluate scientific evidence directly; they consciously often lack the necessary knowledge, time and inclination. Rather, they turn to guidance from those they trust: those they believe are honest and whose values they feel they share.[13] Given that most people are not really in a position to judge the scientific evidence pertaining to the value and risk of GM food, listening to those we trust and perceive to have relevant expertise is a perfectly 'rational' thing to do; in particular, and importantly, we might choose to pay attention to scientific expertise. That, however, is not necessarily the most relevant form of expertise. For many people seeking answers, the most important professionals to turn to might be those with religious, social or ethical authority.

In light of consumers' different decision-making considerations, people eat particular foods for a variety of reasons. For some, food is simply fuel. But food is also, for a large number of consumers, deeply imbued with religious, social, cultural and ethical meanings. So it should be unsurprising that people do not consider what they ingest to be a superficial or trivial decision. As the farmer and author Wendell Berry fondly says, 'eating is an agricultural act.'[14] Moreover, our food choices reflect our values as well as affect our health and our environment. So, listening to those we trust when we recognize the limits of our own knowledge is a pre-eminently wise thing to do. This tendency might be frustrating for those who believe that agriculture and food are simply a matter of scientific progress; people's everyday decisions are rife with examples that are contrary to scientific evidence or scientists' recommendations.

For this and many other reasons, it would be naive to think that the controversies posed by GM food can be easily resolved through scientific education or better public understanding. Just because most people do not possess traditionally defined 'expert' knowledge does not mean that they have no contribution to make to decisions about science and technology. And it definitely does not mean that they should be shut out of the decision-making process. The controversies surrounding GM foods bring a variety of social, political and moral debates to the forefront. Scientists and corporations tamper with our food under the protection of government regulation and in response to the incentives of corporate gain. In this sense, the controversy over GM products reflects ongoing tensions between social and political power, democratic practice and corporate responsibility.

The increasing complexity in global food systems, including the myriad of countries involved in the production, processing and trade of genetically modified foods, amplifies public need for trustworthy and transparent systems. Due to the interdependencies of our ever more complex agricultural and food systems, we must constantly entrust others with our food. And this is no easy thing. Even though many contend that GM food is a wonderful boon to agriculture and is scientifically safe to produce, this does not necessarily reduce public uncertainties. In fact, it adds new ones: people wonder if the process really reflects their values and interests; they are concerned that the scientists, corporations and regulators might not be competent to make the right decisions, and that the wrong decisions will cause harm; and they believe that people with vested interests will communicate overly biased information about the potential risks and benefits. These are just a few of the unknowns that make the public vulnerable to the uncertainties surrounding GM food.

If the industries involved in the production and manufacturing of genetically modified food were involved in a scandal or an event that caused the public to rethink their trust, what would happen? Not much. At least, that has been the case for problems observed thus far. In fact, a United States Government Accountability Office (U.S. GAO) report from 2008 remarked that 'unauthorized releases of GE crops into food, animal feed, or the environment beyond farm fields have occurred, and it is likely that such incidents will occur again.'[15] The report cited six of these events between 2000 and 2008 and suggested strategies to ameliorate the problem and its financial consequences. Yet the problems continue to recur. For example, in 2013 the European Union increased testing and Japan suspended imports of wheat from the United States after an unapproved genetically modified strain of wheat was found growing in a field in Oregon.[16] In 2014 China rejected shipments from the United States containing unapproved genetically modified corn and soy worth U.S.$3 billion.[17] It is tempting to dismiss these two examples as isolated, recent events, claiming simply that accidents sometimes happen. But as political scientist Scott Sagan has remarked about other technical risks, 'things that have never happened before, happen all the time.'[18]

In the end, even when a controversy becomes public knowledge, the public is often without recourse. Now that vast quantities of genetically modified crops are grown and traded all over the world, we generally have to trust our regulators and public watchdogs to do their jobs responsibly and in our best interest, even if experience has shown that these groups can sometimes violate that trust.[19] This is because we are dependent upon numerous manufacturers to provide us with an essential, yet mundane, feature of our lives: food.

Trust becomes the major issue in controversies involving science and technology, especially when the ultimate impact on our lives is not or cannot be settled in advance. People make value-driven choices that cannot simply be reduced to narrow economic or scientific considerations. Given that the important corporate and government stakeholders around the world have chosen to continue progressing with GM food, public efforts to oppose it have become largely limited to fine-tuning its production and, simply put, damage control. In other words, those who either fervently oppose or actively support genetically modified food assume that genetic modification is an intrinsically important technique within the existing global food system. From this perspective, the structures of global agriculture need not be challenged. The 'menu of choice' available to anyone who buys food remains predominantly assembled by experts, scientists, corporations and governments.[20]

All of this implies that, like it or not, genetically modified food will be a part of agricultural practices for some time to come. The open question is how far agribusiness remains dependent on genetic modification. Though GM science has become more evolutionary than revolutionary after thirty years, there are still some scientific unknowns that remain. There have been corporate mistakes, some of which chief stakeholders have admitted. For example, in 1999 Robert Shapiro, then the CEO of the American multinational firm Monsanto, addressed a Greenpeace conference in London via video link to say that Monsanto was going to start listening to its critics, whom it had earlier regarded as 'wrong or at best misguided'. When Monsanto was accused of being a bully trying to force its products on the world, Shapiro remarked, 'If I'm a bully, I don't feel like a very successful bully.'[21] Despite attempts to improve their image, opponents and much of the general public continue to view Monsanto with suspicion and derision. But genetic modification is not *inherently* flawed. Rather, GM crops have been poorly deployed into the real world, with no attempt to fit them into the world's complex social and environmental contexts. Time, energy and money have been dedicated to debates about whether GM food has more potential for good or evil, yet these resources might be better spent seeking solutions to known problems in agricultural practices and systems, like contaminated and inadequate water supplies, degraded soil quality, stresses of climate change and persistent distribution problems. That we have become so focused on genetic modification controversies is the biggest problem of all. The scientific tool of genetic modification is not the ultimate problem, but rather a distraction from the persistent problems that plague our international food system.

Although I have written this book with the primary format of substantial chapters, I attempt to present the controversies surrounding genetically modified food as an unfolding narrative. This demonstrates the back-and-forth nature of the science and public understanding of this issue.

In Chapter One, I explore the complex institutional ecology involved in the production and distribution of food throughout the world. Due to the continued consolidation in the agricultural chain of production, increasing volume of agricultural trade and a complex set of national, regional and international regulatory environments, even issues that appear to be national can have international implications. In Chapter Two, I describe how patents and GM technology completed the transformation of seeds into a commodity controlled by a few firms, creating fear that a few large companies will eventually control the fundamental rights of access to food. Chapter Three endeavours to complicate the assumption that labelling GM food represents a victory over the modern, industrial system of food production. Chapter Four focuses on the contested interests and symbolic battles that characterize scientific judgements and evaluations over claims of expertise, drawing on case studies that reveal a multitude of stakeholders influencing discussions about genetic modification. Chapter Five considers the crucial issue of what, if anything, we have learned as a result of all this commotion over GM foods. I discuss the future of genetically modified food and agriculture as a whole, noting that we are making choices, not taking a predetermined path. The conclusion issues a call to introduce alternative frames with a less distorted vision of the global food system that might lead to better agricultural policy and investment choices. I hope that understanding the social context of the controversies surrounding genetically modified food will play a part in moving beyond the current debates and towards a productive model of agriculture.

1

THE ILLUSION OF DIVERSITY: GLOBAL FOOD PRODUCTION AND DISTRIBUTION

In *Pet Food Politics*, Marion Nestle tells the riveting story of how a few telephone calls about sick cats eventually initiated the largest recall of consumer products in United States history and instigated an international crisis over the safety of imported goods. Nestle tracked the melamine-tainted pet food ingredients along the supply chain to their introduction into feed for pigs, chickens and fish throughout the world. What began as a seemingly small problem for household pets soon signalled a crisis for human health, because of the unexpected connections among the food supplies for pets, farm animals and people.

One outcome of food scares such as this has been the implementation of country-of-origin labelling laws. Although these labels are becoming more common on some foods, processed foods are generally exempted from labelling and multi-country labelling can be problematic. The complexity in this issue reflects the complexity inherent in the global integration of the food supply. For example, a recent study showed that 53 countries contributed to the ingredients of an ordinary chicken Kiev in a Dublin restaurant. The study's scientists used import–export data from the UN and FAO databases to chart the amazingly complex web of networks involved in food transport.[1] Though the complexity and frequency of food trade has many positive implications, the reality involves a number of significant problems. There are inevitable delays in identifying ingredient sources due to the increasingly interwoven nature of the trade network; as a result, in cases of contamination, the number of people placed in jeopardy rises the longer it takes to find the source. Even if countries have become better at preventing food contamination issues, international trade has made it more efficient for the problem to spread. This raises the likelihood of social, political and economic damages with international repercussions. A striking example of this problem can be found in the *E. coli* outbreak of 2011 that was eventually traced to contaminated fenugreek sprouts in Germany and led to illness in eight countries across Europe and North America, resulting in the deaths of 53 people. This outbreak caused more than $1 billion in losses for farmers and industries and hundreds of millions in emergency aid payments to 22 European Union member states.[2] Because of the interconnected web of food trade, the breadth and stringency of domestic and international food safety regulations depends on the cooperation and coordination of government and industry partners, both domestically and internationally.

Similarly, the increased consolidation in agribusiness firms around the world has made international selling more efficient. It should be apparent that agriculture and agrochemicals are global industries, though what is often overlooked by casual observers is how entwined and complex is the system of food production and distribution throughout the world. This system is effectively controlled by a select group of companies. Hundreds of thousands of farmers obtain their seeds and other inputs from this small supply base before food from their crops is made available to billions of consumers around the world. In basic outline, the food production and distribution chain begins when firms like Monsanto, DuPont or Syngenta sell their crop seeds to the farmers who subsequently plant and grow them. These farmers, in turn, sell their crops to grain elevators or handlers such as Archer Daniels Midland, Cargill or Bunge, who sell the grain to food processors. Food and grain processors such as Nestlé and Kraft Foods transform grain into a range of products from bread to cooking oil to snack foods, which are sold to the large food retailers from which most people in developed economies obtain their food, such as Wal-Mart, Tesco or Carrefour.

One analogy to explain the food system was created by Dr William Heffernan and his colleagues at the University of Missouri, who likened it to an hourglass: the grains of sand – in this case, the farm commodities produced on the 570 million farms around the world – must pass through the narrow neck, that is, the few firms that control the processing of the commodities before the food is distributed globally to billions of people.[3] Consequently, the final consumers of GM foods are not the direct customers of the agricultural biotechnology firms. Importantly, this dominance of the food system by just a few companies is a truth obscured from most people due to the numerous choices, which imply a multitude of food suppliers, on retailers' shelves.

At each step in the system, consolidation and globalization have restricted rather than spared agriculture. The severe limitation placed upon farmers is seen in the constraints involved in the purchase of seeds and fertilizer, as well as other inputs, which farmers require every year. Farmers may get these necessities from seed dealers or buy directly from seed companies, but if a farmer wants to purchase seeds genetically modified to be herbicide-tolerant, they will likely also buy a herbicide owned by the same company selling the seed.

Pesticides were a highly profitable business during the 1960s and '70s, but sales had gradually levelled off by the 1990s. The consolidation and control of the agricultural supply chain began in the mid-1970s as agricultural chemical companies began acquiring seed companies, perhaps anti-cipating a time when breeding, genetics and plant molecular biology would replace their agricultural chemicals. The trend in the 1990s of 'life science' companies that tried to create synergy between agriculture and pharmaceuticals research failed to bring significant benefits. Many of these firms spun off the less profitable agricultural chemical side, further driving consolidation in the industry. In 1980, 75 per cent of global pesticide sales were controlled by eighteen companies: Bayer, Ciba-Geigy, Monsanto, Shell, ICI, Rhône-Poulenc, BASF, Eli Lilly, DuPont, Hoechst, Stauffer, Dow, Union Carbide, American Cyanamid, FMC, Rohm & Haas, FBC and Kumiai. By 1985 there were fifteen, and by 1990 there were only twelve: Ciba-Geigy, ICI, Bayer, Rhône-Poulenc, DuPont, DowElanco, Monsanto, Hoechst, BASF, Schering, Sandoz and American Cyanamid.[4] Sales pressure,

increasing research costs and international regulation all
helped fuel industry consolidation and mergers, so that
today just six companies, the 'Big Six' – Monsanto, DuPont
Pioneer, Syngenta, Bayer, Dow and BASF – control 75 per
cent of all private sector plant breeding research, 60 per cent
of the commercial seed market and 76 per cent of global
agricultural chemical sales.[5]

Exploring a timeline of Monsanto's corporate history
provides a spectacular example of the types of acquisitions,
mergers and alliances typical of agribusiness firms.[6] In 1976
Monsanto commercialized glyphosate using the brand name
Roundup. In 1982 Dr Robert Fraley, who would eventually
become the company's executive vice president and chief
technology officer, and his colleagues were among the first
scientists to genetically modify plants.[7] In 1994 Monsanto's
recombinant version of bovine somatotropin (rBST), using
the brand name Posilac, became the first biotechnology
product to win regulatory approval in the United States;
this business was later sold to Eli Lilly in 2008. Monsanto's
history of acquisitions related to plant biotechnology began
in 1996 with the plant biotechnology assets of Agracetus,
the company which had produced the first soybeans with
resistance to Monsanto's Roundup herbicide, along with
several other genetically modified crops; they also bought
out Calgene, the biotechnology company that had developed
the Flavr Savr tomato. Monsanto purchased 40 per cent
of DeKalb Genetics in 1996, allowing it to enter the corn
seed business; two years later Monsanto purchased the
remaining 60 per cent of the company and all of Delta &
Pine Land Company, a producer of cottonseed. Monsanto

also purchased Cargill's seed business in 1998, which gave it access to a broad network of global sales and distribution facilities. In 2005 Monsanto became the world's largest conventional seed company when it purchased Seminis, a leading global vegetable and fruit seed company. Over time, Monsanto acquired several more seed companies, including Jacob Hartz, Agroceres, Asgrow Seed, Holden's Foundation Seed, Limagrain Canada Seeds, Plant Breeding International Cambridge (PBIC), Agroeste Sementes, Western Seed and Poloni Semences.

Though the potential of genetically modified seeds to use less pesticide and create higher yields remains subject to debate, the genetic modification of commodity crops has generated dramatically rapid consolidation of global agri-business. In particular, 'Monsanto has used the huge profits from Roundup Ready seeds to buy up a sizable portion of the seed industry.'[8] One industry analyst has called Monsanto 'the Pac-Man of the industry', because of its propensity to gobble up companies, guaranteeing a large and steady supply of seed to use in its quest for global dominance.[9] Dr Fraley remarked that 'What you're seeing is not just a consolidation of seed companies, it's really a consolidation of the entire food chain.'[10]

Between 2007 and 2010 Monsanto and BASF announced a series of cooperative agreements with total budgets of more than $2 billion focused on the research, development and marketing of new genetically modified plants. The currently proposed Monsanto/Syngenta merger, which would likely result in a new name for the combined firm to reflect its global nature, would mean that one company would be the

world's largest seed company, as well as the world's largest
agricultural chemicals company. This would likely trigger
another wave of consolidation and purchasing through the
industry, with the remaining firms trying to compete with
the combined entity's dominant sales, distribution and
research potential.

The cost of research and development, typically close to
10 per cent of a company's sales, not only spurs consolidation,
but helps the Big Six maintain their leading market positions.
Biotechnology and the discovery of new active ingredients
require long-term, costly research, as does maintaining
a broad range of agrochemical products and meeting
international regulatory standards. The Big Six budget for
research and development is almost $5 billion each year. By
contrast, the budget available for agricultural research in
developing countries is roughly an order of magnitude less,
while the combined budget of the fifteen Research Centers
of the Consultative Group on International Agricultural
Research (CGIAR) is just over $900 million.

As a result, to an astonishing degree, a few dominant
firms set the priorities and research direction of the
modern, emerging global food system. This trend towards
concentration makes sense for much of modern, intermedi-
ate and large-scale farming. Commodity crop farmers buy
seeds, fertilizer, fuel, pesticides and herbicides each season.
These are part of their production costs. Farmers have a
number of options for their seeds. Depending on the method
of production, they can often source them from locally or
regionally based seed dealers or buy them directly from
larger seed companies. However, if a farmer decides to buy

a herbicide-tolerant seed, they are in effect also choosing to buy herbicides owned by the same company that owns the seed genetics. For example, Roundup Ready Soybeans or Corn require the use of Roundup herbicide for farmers to reap the most benefit from the purchase of the GM seed. This is because Roundup Ready Soybeans have been engineered by the Monsanto Corporation to contain an in-plant tolerance to Monsanto's Roundup formulated herbicide, allowing the farmers to spray their crops with this weed killer without harming the crops themselves.

Given the current global agricultural structure, this means that a handful of firms are able to exert abnormal pressure at several stages. This extreme concentration of crop seed production in the hands of a few multinational companies has generated vocal opposition by advocacy organizations, which fear that these companies will control the fundamental right of access to food through their oligopolistic control of seeds and agricultural chemicals. As a limited counterbalance, governments have an interest in preserving competitive markets because these are assumed to provide lower prices and more choice for consumers. Therefore economists and other social science scholars, in addition to government regulators, have become concerned with the extent of this industry's consolidation. Competitors should also worry.

Monsanto's $45 billion attempt to buy Syngenta is instructive. Not only would this acquisition create an agribusiness powerhouse, but it would also prompt major rivals to reconsider strategy and review cross-licensing deals. Most of the Big Six firms have licensing agreements with each other. For example, DuPont and Monsanto have a technology licensing agreement on soybeans; BASF works with Monsanto on herbicides for several crops; Bayer has licensing agreements with each of its peers in seeds technology; and so forth. While it is too early to be certain whether the deal will ultimately go through or what a combined company would look like, it is likely that there will be huge regulatory hurdles involved and a wave of consumer concern.[11] Independently, the two seed companies are dominant, with significant genetic trait businesses and popular pesticides; combined, their similarities will rightfully raise anti-trust concerns. Though the combined entity would have tremendous assets to leverage into new products, those assets could just as easily be used to unduly influence the agribusiness sector; with so much control over such a significant share of the popular seed varieties, genetic traits and chemical inputs, it could dominate the market in an undesirable way. Combined, the single firm would control more than one-third of global seed sales. To put this in perspective in another context, currently the thirty largest food retailers in the world account for one-third of all grocery sales.

Though we could talk about national food systems, agribusiness has truly expanded beyond national borders and become part of the globalized, industrialized food system. The global reach of the dominant firms means that it often makes little sense to speak of the food system of a single country. Even if we were to try to treat a problem in isolation, the sensitive dependence of a global food system can create an inevitable and sometimes unforeseen chain of events around the world. For example, dietary changes and food preferences in one part of the world have implications for production and trade in other parts. Less and less food that is produced locally is actually consumed locally. Instead, local products get traded and transported elsewhere, either as ingredients or sellable foods, resulting in more than $1 trillion in global food trade each year. This increasing inter-dependency means that changes in one country's pattern of consumption can have major impacts on food markets throughout the world. For example, as rising incomes have increased demand for meat in China, livestock producers have increased their need for imported corn, a major component of animal feed.

The continued consolidation in the agricultural chain of production, increasing volume of agricultural trade, and a set of complex national, regional and international regulatory environments complicate international trade in genetically modified food products. Many countries have handed governmental regulatory responsibility for GM food to multiple agencies that deal with agriculture, the environment and food safety. These agencies have then typically grafted regulations concerning GM food and crops onto existing regulations related to the release of new varieties, use of pesticides and the marketing of food products. While developed countries have focused on domestic priorities and strategies when establishing their regulations to deal with GM crops, most developing countries are more constrained by the requests and expectations of their main trade partners. For developing countries, reconciling their trade interests with their responsibility to their own citizenry is an increasingly complex task, and one that is not new to agriculture.

Farmers' incomes have been dependent on international markets for fifty years. More recently, global and regional trade agreements such as the General Agreement on Tariffs and Trade (GATT), the North American Free Trade Agreement (NAFTA), the Canada and European Union Comprehensive Economic and Trade Agreement (CETA), the Trans-Pacific Partnership (TPP) and the Transatlantic Trade and Investment Partnership (TTIP) have been the focal point of much of the globalization discussion. The expansion of agricultural market access is an important dimension of international trade. However, perhaps the most important feature of these trade agreements is the movement to large-scale production techniques that use standardized technology and management. Standardization is important for lowering production costs and for producing more uniform crops that fit processor specifications, which helps with food safety concerns and can therefore successfully meet consumers' needs. Such technological advances combined with continued pressures to control costs and improve quality are expected to provide incentives for further standardization and industrialization of agriculture.

This more industrialized form of agriculture profoundly changes the competitive environment in the industry. The largest agribusiness firms used to control a limited range of commodities, such as sugar, and a step or two in the processing chain; now, the industrial control of agriculture is concentrated at historically unparalleled levels.[12] The formerly distinct and diverse sectors of the agriculture and food system have become enmeshed through mergers, acquisitions, licensing agreements and all manner of alliances. By expanding their influence and integrating previously disparate points on the same production path, the Big Six agribusiness firms leverage control in the global food chain. Though activists talk about 'farm-to-fork', 'dirt-to-dinner' and 'plough-to-plate' land stewardship and ethical eating, in terms of industry control perhaps the more appropriate phrasing would be 'gene to supermarket shelf'.[13] Controlling so many steps of the food chain means these firms exercise an inordinate and largely invisible level of influence over global decision-making about food. In the last twenty years the global agricultural food market has moved from being more or less competitive to oligopolistic. As food safety experts have discovered, it is almost impossible to ascertain the true origins of any given foodstuff, even if health and illness are on the line. So even though there are fewer and fewer firms involved in all aspects of the agricultural system, the path through the food system has become a maddening, impenetrable maze, even to experts. It has, however, helped produce a relatively varied, inexpensive and plentiful supply of food. That is not to say that access to the food is evenly distributed. Hunger is not a random condition: women, children, indigenous people and other minorities are disproportionately represented among the world's hungry. This is because these groups often have unequal access to resources and participate less in decision-making.

Advocates of genetic modification focus on its potential to increase crop yield, without questioning its consequences throughout the food chain. They typically assert that genetically modified food will feed the world and therefore the huge investments in research and development of this technology will be worth it. Similarly, in recent years many have raised fears that the world may not be able to grow enough food and other commodities to ensure that future populations are adequately fed. Though global food shortages are unlikely, serious problems already exist at national and local levels, and may worsen unless focused efforts are made. Contrary to many who espouse increasing the size of the existing agribusiness firms to achieve greater economies of scale, Dr Hilal Elver, the United Nations' Special Rapporteur on the Right to Food, believes that governments should shift their subsidies and research funding priorities to small-scale farmers.[14] After all, these farmers already feed the majority of the world. Globally, we produce almost 20 per cent more food per person than just three decades ago.[15] Currently, the number of food markets and ways people can access them is inadequate. This is especially true in rural areas and developing countries. Moreover, in these areas there are few high-quality road transit systems to transport crops, producers and consumers to the centralized markets. In places with full access, many people are too poor to afford enough food to feed themselves and their families. This is also true for farmers, who cannot secure an adequate payment for their crops that allows them to pay off their costs of production, including money spent on seeds, fertilizers, pesticides, herbicides and fuel. Indeed, given its potential to displace large numbers of people and to deny them access to their means of subsistence, the imposition of genetically modified food practices could be detrimental to helping some people throughout the world to feed themselves.[16]

Increasing market share and control of the food system has become the guiding principle of the modern biotechnological revolution in agriculture. Though often conflated with this, the sheer size of the agribusiness firms is not the principal concern. The concern instead is that as fewer input suppliers increasingly dominate the market, farmers' choices will become more restricted, resulting in only a few products that will be pushed as the industry standard, therefore promoting a narrower genetic base for agriculture as well as restricting the type of farm enterprises. In particular, given that the industry favours technological solutions to farming problems, the choices offered to farmers would likely mean trying increasingly advanced chemical, biological and genetic technologies to solve new problems that will inevitably arise. Detractors fear that this ongoing sequence will end up like the nursery rhyme about the old lady who swallowed a fly: 'She swallowed the dog to catch the cat, she swallowed the cat to catch the bird, she swallowed the bird to catch the spider . . . she swallowed the spider to catch the fly, I don't know why she swallowed the fly, perhaps she'll die.'

Farmers, however, are not necessarily as concerned by the seemingly inevitable treadmill. Genetically modified seeds can reduce labour and enable a more flexible herbicide or pesticide routine. It is easy to see the appeal, given the potential for time and labour savings. Moreover, farmers who have stayed in business have already adapted to new technologies and use them on their farms, so they have a high degree of confidence in science and technological innovations.[17] As the International Service for the Acquisition of Agri-biotech Applications (ISAAA) is fond of pointing out, farmers have increased their planting of genetically modified crops every year since 1996.[18] They assert that this trend reflects the growing and continued confidence of farmers around the world. But this agricultural and economic trend causes a cascade of effects.

A strong case can be made that there are now fewer breeding programmes for non-GM seeds and that it is becoming increasingly difficult to keep non-GM seeds pure.[19] In countries such as the United States that do not heavily regulate or restrict access to genetically modified crop varieties, the increased planting of GM crops eventually leads to fewer non-GM seed choices or to choices that are restricted to GM cultivars only.[20] Similar observations have been reported in India and South Africa.[21] Global corporations like DuPont, Monsanto and BASF who produce genetically modified seeds tend to offer fewer non-GM varieties. Private seed companies seem to limit the choices they offer in GM-adopting countries to GM seeds under patent protection; in contrast, farmers in non-GM-adopting nations continue to have the same or increased choice of seed cultivars since the introduction of GM seeds in the global market.[22] Unlike the dominant private seed companies, regionally and locally based seed firms contribute significantly to the numbers of seed cultivars in non-GM-adopting Europe. In the limited research that has been completed, it has been shown that seed options narrow when a handful of seed companies that offer genetically modified seeds dominate the marketplace. So rather than farmers having more choice, the available evidence points to more restricted options for farmers over time. Still, given limited research and several plausible, and contradictory, outcomes, there should be more thought given to how the increased market concentration of seed suppliers affects crop seed diversity, seed prices and farmers' planting decisions and options.[23]

The rapid adoption of GM crop seeds is a classic example of a technology treadmill in action. These effects extend to all types and sizes of farms. The site of production has increasingly moved from family-based, small-scale, relatively independent farms to larger farms that are more closely aligned with agribusiness firms across the production and distribution chains. The concentration and integration of the agricultural inputs and processing sectors profoundly speed up the technology treadmill for farmers.[24] As the theory suggests, as more farmers get on the treadmill and adopt the technology, overall production goes up while prices go down. This means that profits are no longer possible, even with lower production costs. Consequently, increasingly new technology must be adopted to make profits possible again. Confronted with the rapidly expanding production technologies, farmers are forced to either exit the industry or remain competitive by becoming loyal to the pursuit of ever-newer technology. The more successful farms absorb the struggling ones that are no longer competitive, leaving behind a structure of a few larger farms. As a result of this technology-driven increase in agricultural productivity, farmers are subject to a cascade of effects along a technocratic, bureaucratic path that is more comfortably travelled by modern multinational agribusiness firms than any other form of organization.

At this point we should keep in mind the American psychologist Abraham H. Maslow's golden hammer metaphor for narrow professional thinking: 'if all you have is a hammer, everything looks like a nail.'[25] In context, if we think about problems of yield, there are a number of ways to approach them. One approach, the one we most commonly follow, would be to apply scientific insights into plant genetics to create seeds that will produce desired qualities and improve farmers' chances of producing the yields they want. This view privileges the scientific expertise of companies with investments in chemical, biological and genetic resources. Moreover, because of regular scientific advances, companies continue to establish their intellectual property portfolios, further entrenching their expert status.

But researching agricultural problems, even fundamental tenets like crop yields, can require expertise in a wide range of fields – such as plant breeding, genomics, soil chemistry, irrigation, agricultural economics or pest management – each of which has distinct values, customs and standards of evidence. This means that achieving scientific consensus on which is the best course of action to follow can be challenging. And that is just the science. Imagine the kinds of expertise and costs that would be necessary if agribusiness firms decided that data analytic improvements would increase crop yields; rather than, or in addition to, investing in breeding programmes, it would make sense to fund a broadband data infrastructure for farmers. However, this would not draw on the traditional strengths of agribusiness firms and it would be surprising for them to embrace this mission. As Thorstein Veblen wrote: 'What is consistent with the habitual course of action is consistent with the habitual line of thought, and gives the definitive ground of knowledge as well as the conventional standard of complacency or approval in any community.'[26] Though necessary for developing genomic solutions that have produced GMOs, the trained incapacity of agribusiness firms can blind them to solutions that do not take advantage of their expertise. The industry has spent billions of dollars shepherding GM foods to market and they will not toss aside their investment soon or easily. However, there are other paths to take.

No matter which road they choose, companies must take into account the social, cultural, religious, ethical, economic, legal and political contexts. It is no wonder that working on issues surrounding food can be exasperating. Just trying to understand and reconcile even seemingly simple questions puts everyone at the centre of an interdisciplinary maelstrom. No matter how strong the scientific insight, it is worthless if blind to the broader context. I do not intend this to be a divisive statement, casting one view as right and one view as wrong. This also does not mean that scientific discussion necessarily devolves into disorder. Rather, I wish to enlarge our conception of potential solutions to be more inclusive of varying social, cultural and ethical contexts. Whether industry can make the right decision depends on your definition of right. In the end, the public's determination is unlikely to match that of the experts.

The advancement of genetic modification has become inextricably entwined with agribusiness concentration and industrialized farming. So if market concentration is socially problematic, then genetically modified crops will be problematic because they are a particularly visible marker of modern industrial farming practices. In the abstract, genetic modification need not be more problematic than any other agricultural tools and technologies that have enabled industry consolidation. The complicated mix of subsidies, incentives, insurance programmes, grain reserves, food entitlements and credit schemes all have so many moving parts that they are very difficult to analyse. In this milieu, food policy choices are often extremely political and, in the age of globalization, such policies have become more and more concerned with trade rather than with national production. The growing contribution of trade and foreign direct investment in determining the types of inputs into agriculture, into the structure of food markets and in the globalization of the food industry has impacted all dimensions of the food system. The common feature for each of these issues is that the concentration of agribusiness firms is strongly linked to their power over consumer choice. The Big Six firms' influence over the food system is now remarkable, whether one looks nationally, regionally or globally. As a result, examining the concentration of corporate power should be the first step in trying to determine who will feed us and what we will eat.

This assessment is not necessarily bleak. Though the Big Six steer future research, they receive feedback from a wide array of contributors that exert some influence in the global food system, including farmers, processors, policy-makers and consumers. Cost is not necessarily the foremost priority for each of these; they have diverse goals that include improving labour conditions, public health, environmental preservation, biodiversity protection and increasing productivity. In the end, consumer purchases signal what and how much to produce to maintain a relatively varied, relatively inexpensive and plentiful supply of food. Moreover, interconnectedness means that trust is a central issue in food policy. The focus on trust should remind us that consumers have played, and can continue to play, an important part in the evolution of food policy. Methods of food production, distribution and consumption can be subjected to considerable debate and critical appraisal. Agribusiness will listen.

The relationship between industry and consumers is bounded, however. Even if large groups of consumers demanded it, in a unified voice, it would be historically myopic to pronounce an end to the industrialized system. Rhetoric and public action campaigns suggest that the food chain is consumer-led, but this is not entirely accurate. Dr Jean Kinsey, former president of the American Agricultural Economics Association (AAEA), has argued that the old supply–demand chain is a now a loop where intelligence is gathered about consumers but shaped by requirements coming back up the supply chain. This means that consumers can force decision-makers to alter part of the food system to better reflect their priorities. In short, the food system is complex but adaptable. Given the extreme interconnectedness, interactions between part of the food system cannot be fully understood in isolation. Studies to inform food and agricultural decisions, therefore, require an analytical approach and methodologies capable of considering the full range of key interactions, adaptations and other features of complex systems.

2
INTELLECTUAL PROPERTY: PROTECTING OR OVERREACHING?

The concentration of crop seed production in the hands of a few multinational companies has given rise to the fear that a few large companies will control the supply of seeds and food and may eventually control the fundamental rights of access to food. GM crop seeds are patented, with seeds leased to farmers on an annual basis. As multinational agribusiness firms consolidate their presence in the international market, real choice for farmers may evaporate; farmers can become locked into a system in which they have little or no choice over what to grow and with which chemicals, who to sell to and at what price. The regulatory infrastructure created by, and enforced with, patent law gives the companies that control the technology a uniquely privileged position and great influence over farming practices. The enforcement of intellectual property rights in this area leads to the fear that the adoption of GM crops will transfer resources from the public sphere to private ownership. As a result, the World Trade Organization's controversial Trade Related Aspects of Intellectual Property Rights (TRIPs) agreement, which requires nations to establish some form of protection for plant varieties, is the focus of international scrutiny.

Such considerations are not limited to the developing world. Some u.s. farmers have also been taken by surprise by prohibitions on replanting or reselling seed, and some find this an economic hardship and an intrusion on what they see as their historical privilege – not such an extreme economic hardship as would be experienced by subsistence farmers across less developed nations, but an economic hardship nonetheless. The assumption is that subsistence farmers in the developing world are the most vulnerable to new patent restrictions, but wherever this kind of complex impact on the global food system occurs, its significance – just like the significance of the science and technology that produced GM alternatives in the first place – is difficult for even experts to evaluate. It is not impossible to determine, but it is not a simple or straightforward problem with a single solution. Moreover, modern biotechnology pushes legal and scientific debates to new arenas, and veers into the practical boundaries of philosophy, morality and ethics.

Article 27 of the Universal Declaration of Human Rights recognizes the 'right to the protection of the moral and material interests resulting from any scientific, literary or artistic production'.[1] Simply put, intellectual property rights say that inventors and owners of a product should benefit from their invention. This has been legally recognized at least as far back as the Paris Convention for the Protection of Industrial Property of 1883 and the Berne Convention for the Protection of Literary and Artistic Works in 1886. But what exactly do we mean by inventor, owner and product? Moreover, what effect will the patenting of scientific discoveries likely engender? When dealing with biological science,

where is the boundary between natural and man-made?
In other words, where does science begin and where does
nature end? Is it really fair if the inventor is the only one who
should benefit? In his *Two Treatises of Government*, originally
published in 1689, John Locke wrote:

> As justice gives every man a title to the product of
> his honest industry, and the fair acquisitions of his
> ancestors descended to him; so charity gives every man
> a title to so much out of another's plenty as will keep
> him from extreme want . . .[2]

In this tradition, justice and charity are provided as reasonable claims to property. This means that the inventor should
not be singled out for praise and remuneration. Instead, the
recipients of charity are equally deserving of dignity and
consideration. So, in practice, how does intellectual property
balance 'the moral and economic rights of creators and
inventors with the wider interests and needs of the society'?[3]

To some, the crux of the matter regarding genetically
modified food rests in the arcane, contested and rapidly
evolving world of international intellectual property rights.
In this realm the Indian environmental activist and anti-
globalization author Vandana Shiva is a polarizing figure,
seen by some as a saviour and by others as a charlatan.[4] She
has argued for more than 25 years against corporate agricultural and food practices relating to intellectual property,
biodiversity and biotechnology. The non-governmental
organization she founded, Navdanya, promotes small, bio-
diverse and organic farming, native seed saving and fair

trade as a way to thwart corporate attempts to patent seeds, crops or life forms. Though it is not terribly controversial to say that incentives to copy and reproduce seeds fundamentally changed as agriculture evolved into an industry as much as an act of subsistence, is it fair to claim, as Shiva does, that international patent law reflects the arrogance of Western civilization?[5] Alternatively, could you argue that patent protections are a net societal good and unlock corporate creativity?

Unique knowledge and specialized understanding are features of modern economic and social life. The effects of patents, both good and ill, are based on governments' grants of limited monopolies to inventors. Intellectual property and patent laws try to balance the tension between the desire to spread knowledge and the potential for economic gain from that knowledge. Proponents argue that corporations would not want to invest time, energy and capital into new technologies without government protections because they would not be able to recoup their costs or realize a profit from their new inventions. Opponents argue that patents allow established companies to maintain their market positions and limit the competition that would otherwise spark innovation. These debates are not entirely new; patents have been a feature of agriculture for close to a century.

The u.s. Plant Patent Act of 1930 gave patent protection rights to the developers of asexually reproduced plants such as apple trees and rose bushes, which are propagated by cutting pieces of the stem rather than by germinating seed. The u.s. Plant Variety Protection Act of 1970 established patent rights for developers of new varieties of seed-propagated plants, but specifically exempted F1 hybrids, including carrots, celery, cucumbers, okra, peppers and tomatoes, from patent protection.[6] Farmers, however, were granted an exemption allowing them to save patented seeds for use on their own farms and permitting them to sell patented seeds to other farmers. Matching the protections in the International Union for the Protection of New Varieties of Plants (UPOV) Convention of 1961, U.S. plant breeders were granted an exemption to use patented plants to develop new patentable varieties.

Private firms are not the only ones who request and receive patent protection. Public and private universities, the U.S. Department of Agriculture and other similar international agencies often hold plant variety patents. For example, in the early 1970s Stanley Cohen of Stanford University and Herbert Boyer of the University of California at San Francisco developed a laboratory process that produced recombinant DNA products. The process was revolutionary for molecular biology. Despite its potential for tremendous commercial success, it was also at the centre of vigorous debates. Initial concerns about the social and environmental implications gave way to the debate about who should own, control and profit from scientific knowledge.[7] In 1974 Stanford University and the University of California applied for a patent on the Boyer–Cohen recombinant DNA process. In June 1980 the *Diamond* v. *Chakrabarty* U.S. Supreme Court decision, written by Chief Justice Warren E. Burger, deemed that patentable subject-matter included 'anything under the sun that is made by man'.[8] This controversial decision upheld the patent on a genetically

modified bacterium for digesting crude oil in oil spills. The
U.S. Patent and Trademark Office had originally rejected
the patent, but Dr Ananda Mohan Chakrabarty, an Indian
American microbiologist who developed a genetically engineered organism, a new species of oil-metabolizing bacteria,
using plasmid transfer while working at General Electric,
appealed to the U.S. Supreme Court. This five-to-four Court
ruling decreed the patent eligibility of both transgenic microorganisms and DNA, making at least some biotechnological
inventions patent-eligible, and cleared the way for the first of
three Boyer–Cohen patent applications, which was approved
in December 1980. The *ex parte Hibberd* ruling in 1985
expanded utility patent protections to the methods used to
engineer a plant, including genetic sequences inserted into
a species' genome and the resulting plant.[9] These changes
made U.S. patent laws similar to those in effect in Europe
under the 1991 UPOV Convention.

Extending patentability in this manner created a new set
of intellectual property rights and ownership claims. The
nascent biotechnology industry leveraged their newfound
patent rights to acquire financing and to protect their
investments during the long commercialization process.
This liberal patentability standard was adopted worldwide
in Article 27 of the TRIPS agreement that entered into force
in 1995. As a result, the intellectual property surrounding
the genetic modification of seeds and plants became commodities with their own exchange value. The Boyer–Cohen
patents, in force until 1997, were used by 468 companies
and yielded more than $250 million in licensing revenues
during their lifespan.[10] Rather than only maximizing

profitability, these patents established a de facto exemption for academic research institutions, meaning licence fees were based on commercial end-use products rather than non-commercial scientific work consistent with the norms of open science and permitted by Article 30 of the TRIPS. These changes to patent law created and transformed commercialized biotechnology as much as the scientific breakthrough itself.[11] In many ways, the debates around this patent were a harbinger of the social and ethical issues associated with GMOs.

By the time biotechnology had been embraced in the agricultural seed market in the 1990s, plant and seed patents were commonplace. But being able to patent a genetic modification technique or gene sequence is a far cry from commercializing a genetically modified seed. By itself, one patent means very little. Though often seen as necessary, it is far from sufficient. Agribusiness and biotechnology firms often control or hold hundreds of patents; arranging to license these existing steps is often a necessary precondition for bringing a desired trait to market. If a company successfully develops a new trait, it can commercialize it in its own seeds or license it more broadly – this can create a complicated web of intellectual property, even for a single seed crop.

The Big Six seed, biotechnology and agrochemical firms – BASF, Bayer, Dow Agrosciences, DuPont Pioneer, Monsanto and Syngenta – have entered into a number of agreements to share patented, genetically modified seed traits, such as herbicide tolerance and expression of insecticidal toxins.[12] In fact, when Monsanto filed a lawsuit against DuPont for patent infringement related to its soybean herbicide, their settlement resulted in a long-term, billion-dollar licensing agreement between the two companies. These companies prefer carefully crafted licensing agreements because of the time and money required to bring a new plant biotechnology trait to market. In 2011 the agribusiness consulting firm Phillips McDougall used data provided by the Big Six to determine the relative cost and duration of this process. They found that the discovery, development and authorization of a new plant biotechnology trait took anywhere from 11.7 to 16.3 years and cost $136 million.[13] These costs and timelines are likely conservative estimates when it comes to approval for the more recent stacked trait varieties that are the final product in most crops today.

Plant patent laws and multinational patent agreements are the foundation of the modern biotechnology-based agribusiness food system that the anti-GMO movement denounces. This is because the patent laws act as the legal framework that encourages agribusinesses like Monsanto to amass their global seed and insecticide portfolio. This, in practice, lets these agribusiness firms establish a vertical monopoly: to get the company's highest-yielding seeds, you will need to use the company's agricultural chemicals; if you want to use the company's chemicals, you will want to buy their seeds, which are optimized to thrive when used with those specific chemicals. This is not a simple case of 'freebie marketing' where one product is sold at a discount and the second, dependent product is sold at a considerably higher price (this is often called the 'razor blade' business model, because razors can be very inexpensive but the required replacement blades can be costly; similarly, inkjet and laser printers are often sold at or below the manufacturing cost to generate sales of proprietary cartridges that will generate greater profits over the life of the equipment). The vertical integration of agribusiness firms such as Monsanto means that the firms may sell patented seeds and a corresponding chemical input, while licensing patented traits to other seed companies that in turn offer GM seeds. This means that you can buy the entire bundle from Monsanto, or buy from another company that has already licensed the technology from Monsanto. As a result, agribusiness firms' GM seeds compete for seed sales against independent seed firms licensing the same GM traits. The extent to which and how licensing arrangements alter the competitive seed market is an emerging issue.[14]

The extremely large financial costs and long timeframes for patent approvals have set up the basis for a patent oligopoly at the global level. Only wealthy and stable firms can afford to compete, and government-granted patent rights assure that other competition is restricted. Though in many ways patents can help make the already rich even richer, their patents do not grant them unlimited power, since they have a finite term. For example, Monsanto's patent on glyphosate, the active ingredient in the herbicide Roundup, expired in 2000 and its patent on the first generation of Roundup Ready seeds expired in 2015.[15] To stay ahead of the competition, agribusiness firms are forced to innovate prior to the expiration of the protections their patents afford them. This means they have to continually investigate any promising pathways so that they can maintain their market share. Prior to the expiration of its first-generation patents,

Monsanto had already launched its Roundup Ready 2 Yield products, complete with new and longer-lasting patent protections, as the base for its future biotech traits.[16]

Patent expiration also opens opportunities for alternatives. For example, the University of Arkansas System Division of Agriculture has released its first soybean variety that features Roundup Ready technology. Unlike Roundup Ready seeds from Monsanto, the University of Arkansas seeds do not have added technology fees that have traditionally been assessed to profit from intellectual property arrangements for plant breeding. Moreover, growers are now allowed to save seed from each harvest for planting the following year. These seeds will not present a financial challenge to Monsanto, but they point to the possibility of alternative intellectual property.

The Open Source Seed Initiative (OSSI), founded in 2012, is one such alternative. OSSI intended to stimulate innovation in plant breeding by creating a licensing framework similar to that used by the open source software movement.[17] Rather than becoming a seed company or even a distributor of seeds, they wanted to develop a legally defensible licence for germplasm that would preserve the right to the un-encumbered use of shared seeds and their progeny. But after two years of work, OSSI ran into two big problems.[18] First, the complex legal requirements of the licence made its transmission impractical and unlikely. Second, the idea of a legal licence was philosophically and politically unacceptable to many organizations, such as La Vía Campesina and Navdanya, that had a shared goal of seed sovereignty. Consequently in 2014 OSSI decided to abandon a legal licence and instead shifted their focus to a pledge that encapsulated their norms and ethics:

> You have the freedom to use these ossi-Pledged seeds in any way you choose. In return, you pledge not to restrict others' use of these seeds or their derivatives by patents, licenses or other means, and to include this pledge with any transfer of these seeds or their derivatives.[19]

Although this phrasing would not withstand any legal challenge, it represents an uncompromising commitment to free exchange and use. It is also a potentially effective tool for outreach and education that could get people thinking and talking about how seeds are controlled.

Moreover, agribusiness firms and university patent holders have, in a few notable instances, seen fit to relinquish their patents. The Rainbow papaya, an F1 hybrid variety of the standard yellow-flesh export variety papaya Kapoho Solo crossed with the red-flesh SunUp, offers a remarkable example. Researchers from Cornell University and the University of Hawaii had worked since 1984 to develop this Rainbow variety, which includes a gene that makes the papaya plants resistant to the ringspot virus – an infection that decimated Hawaiian papaya production by more than 50 per cent in the 1990s. Dennis Gonsalves, then a plant scientist at Cornell University and now the director of the USDA's Pacific Basin Agricultural Research Center, used genetic engineering to insert a small portion of the ringspot virus into papaya genetic material, creating the Rainbow papaya. The USDA, the

Environmental Protection Agency (EPA) and the Food and Drug Administration reviewed and approved the genetically modified papaya for production and sale in September 1997. After the regulatory review, Hawaii's Papaya Administrative Committee successfully negotiated licence agreements with all the owners of the patented genetic engineering technology that went into the creation of the Rainbow papaya. In the end, this involved agreements with four companies: Monsanto, Asgrow Seed Company, Cambia Biosystems LLC and the Massachusetts Institute of Technology.[20] Gonsalves and the Hawaiian public sector introduced this genetically modified papaya – which proved resistant to ringspot virus attacks – to the market, and on 1 May 1998, after the patent licences came through, Hawaii's Papaya Administrative Committee distributed Rainbow seeds for free to the Big Island's growers.[21] Though not entirely free of controversy, the Hawaiian papaya crop is fairly stable now, and more than 75 per cent of the island's papayas are the genetically modified Rainbow variety. Moreover, these papayas are the only genetically modified fruit freely grown in the United States.

Though important to Hawaiian farmers, papaya is on the global scale only a minor crop. While agribusiness firms have concentrated their efforts on big market commodity crops, interest in improving local or speciality fruits and vegetables has faltered; the industry chooses to pursue greater profits than can be had with these speciality crops. Given that many of the patents are held by seed companies with a preference for blockbuster profits, the miniscule profit potential of speciality crops means that they are unlikely to be developed. Even if an agribusiness firm were to forgo the possibility of large-scale profits, scientists would still need to spend multiple years of their careers and millions of dollars in legal costs to gain the necessary regulatory approvals. Moreover, there is relatively little grant money or academic prestige offered to applied scientists building on the pioneering work of their predecessors. As a result, most small speciality crops have become orphan crops, lacking consistent research and development by scientists familiar with biotechnology.[22] In its evaluation of biotech crops, the U.S. National Academy of Sciences cited the lack of biotech work on speciality crops as one of farming's most pressing problems.[23]

There has been some development more recently, however. Florida citrus growers have collectively financed research into a genetically modified greening-resistant tree.[24] Okanagan, a small Canadian firm, also won USDA approval for their genetically modified apple trees, which produce fruit that resists turning brown when sliced or bruised.[25]

In contrast to the small profile of speciality crops like papaya, rice is one of the world's primary staple crops. More than 3.5 billion people depend on rice for more than 20 per cent of their daily calories. It has been cultivated for thousands of years and is particularly important in Asia, but is also important in Latin America and parts of Africa. 'Golden Rice' is rice that scientists have genetically modified to produce and accumulate beta-carotene in its endosperm, the starchy interior part that is all most people eat. The increased beta-carotene gives the grains a light yellow colour as opposed to ordinary white rice. Golden Rice could theoretically contribute to a reduction of chronic health problems caused by vitamin A deficiency, because the beta-carotene from the rice would be either stored in the body's fatty tissues or converted to vitamin A. Vitamin A deficiency is a serious public health problem, especially in Africa and Southeast Asia. Not only is the vitamin crucial for maternal and child survival during pregnancy, according to the World Health Organization, the lack of vitamin A 'causes a needlessly high risk of disease and death'. Globally, one-fifth of children under five years old, roughly 127 million children, are vitamin A deficient; almost one-quarter of early childhood deaths due to measles, diarrhoea and malaria are attributable to vitamin A deficiency, accounting for more than 600,000 deaths of preschool-age children every year.[26] Clearly this is a problem that, if solved, could greatly improve the health and well-being of large swathes of the global population.

Ingo Potrykus, a professor of plant sciences at the Swiss Federal Institute of Technology in Zürich, along with Peter Beyer, a professor of cell biology at the University of Freiburg, Germany, are considered the inventors of Golden Rice. Though rice had been a target of biotechnology research in the 1980s, Potrykus and Beyer started their work together in the early 1990s, and in January 2000 *Science* published the results of their research, heralding the start of Golden Rice.[27] This rice project represented a considerable technical advancement in genetic modification, but it soon became embroiled in public debate.

Golden Rice first made mainstream headlines in July 2000 when the cover of *Time* magazine boldly proclaimed: 'This rice could save a million kids a year.'[28] While supporters argue that it will save lives, opponents have raised concerns about threats to biodiversity, potential health and safety risks, and the need for more scientific data and trials to support the viability of the project. Critics such as Vandana Shiva have dismissed the development process entirely, contending that it is a hoax intended as a Trojan horse to gain public support for genetically modified crops of all sorts that would benefit global agribusiness at the expense of consumers and poor farmers.[29] Similarly, the influential author Michael Pollan suggested that it marked the advent 'of the world's first purely rhetorical technology' that would offer more in public relations to beleaguered agribusiness firms than it ever would to help under- and malnourished children.[30]

Funding for the research came from several sources – a fact that complicated intellectual property issues. Due to Potrykus and Beyer's uncertainty about which intellectual property rights they had used in developing their enhanced rice, the ISAAA conducted an audit to find out how it could be released for humanitarian purposes without infringing on intellectual property rights. Astonishingly, some seventy intellectual property rights and technical property rights belonging to 32 different companies and universities had been used in the experiments.[31] Even though private companies holding the patents agreed to make their intellectual property available on a royalty-free basis in developing countries, the process of bringing this project to commercial completion has been legally complicated. Currently, the Golden Rice Humanitarian Board has the right to sublicense the technology to breeding institutions in developing countries, free of charge.[32] Furthermore, public sector institutions in Bangladesh, China, India, Indonesia, the Philippines and

Vietnam are attempting to develop local Golden Rice varieties. It has yet to meet its own stated goals, however; fifteen years after the *Time* magazine cover story, Golden Rice is still not available to farmers or consumers. If and when Golden Rice will ever make it to market is unknown. Its much-hyped potential, as well as the potential for patent problems, has run far ahead of reality.

Concerns about intellectual property and patents also exist on a smaller scale, and Monsanto and other agribusiness firms have faced criticism for being overly aggressive towards farmers in their use of patents and intellectual property to safeguard their investment in biotechnology. Perhaps the most notorious case occurred in 1998, with Percy Schmeiser, a 73-year-old farmer in Saskatchewan, Canada, who had not paid a licence fee but grew Monsanto's genetically modified Roundup Ready rapeseed (canola), which he claimed had accidentally reached his farm via cross-pollination or accidental contamination from neighbouring farms and roads. Schmeiser claimed that he did not knowingly acquire the technology or segregate the contaminated seeds, or spray his crop with Roundup. The Court stated that it could not be sure of the rapeseed's origin and that the rapeseed plants may have come from Roundup Ready seed that had drifted onto Schmeiser's farm, but nevertheless the Court argued that this was not a reasonable explanation given that more than 95 per cent of his fields were planted with Roundup Ready seeds. Monsanto's allegation of patent infringement was upheld in a five-to-four split decision in Canada's Supreme Court. Though the Court ruled that Schmeiser's unlicensed use of seed containing Monsanto's patented gene was sufficient to constitute infringement, it awarded no damages on the grounds that the farmer had not economically benefited.[33] This meant that Schmeiser would have to pay to Monsanto neither the $19,000 in estimated profits on his 1998 crop, nor Monsanto's legal costs estimated at more than $150,000. Given that Monsanto won this appeal, many have cited this landmark case as a warning that patent protection can prevail over landowner rights and that farmers might have to guard against the presence of genetically modified seeds lest overzealous agribusiness bullies sue them. On the other hand, the narrow ruling could be seen as a warning to GM seed producers that they would have a difficult time collecting damages for patent infringement and could face unlimited liability if a farmer, perhaps one with a certified GM-free organic farm, could establish damage.

More recently, the Organic Seed Growers and Trade Association, along with more than sixty agricultural organizations, sued Monsanto in 2011. The suit sought to proactively prohibit Monsanto from suing farmers whose fields become inadvertently contaminated with the company's patented crop seeds. In the suit filed by the organic and conventional growers, the group asked Monsanto for a pre-emptive pledge not to sue them since they do not use the genetically modified seeds. However, Monsanto has refused to sign a pre-emptive pledge of this nature, claiming that it would enable intentional infringement on their patents. The lawsuit against Monsanto was dismissed in 2012 in district court, a decision upheld in 2013 by the U.S. Court of Appeals for the Federal Circuit. The complainants petitioned the U.S. Supreme Court to hear their case, but in 2014 the Supreme Court declined the opportunity to review the lower court's rulings.[34]

In another challenge, the 2013 U.S. Supreme Court case *Bowman* v. *Monsanto Co. et al.*, the judges unanimously ruled that farmers could not use Monsanto's patented seeds to create new seeds without paying the company a fee.[35] This case involved Vernon Hugh Bowman, a 75-year-old Indiana soybean farmer, whose normal farming practice included planting Monsanto's Roundup Ready seeds for his main soybean crop and signing the standard agreement not to save any of his seed and replant it for the next year. For a second harvest he decided to buy some leftover seeds from a small, local grain elevator; these seeds are normally sold for animal feed, food processing or industrial uses. Hoping they were Roundup Ready, Bowman planted them and sprayed them with glyphosate; he then replanted seeds from the surviving plants. Resistance to glyphosate confirmed they were Roundup Ready seeds, which are resistant to the herbicide. Bowman argued that the doctrine of patent exhaustion allowed him to do what he liked with seeds he had legally obtained from the grain elevator. The Court ruled that Bowman could legally do several things with the seed, including resell them, consume them or feed them to animals. They also ruled that by replanting successive generations of the seed, Bowman had, in effect, made additional patented soybeans without Monsanto's permission. The Court ordered Bowman to pay more than $84,000 for infringing on Monsanto's patent rights.

For individual farmers as well as for nation-states, intellectual property and patent laws for genetically modified seeds can be complicated. Simple explanations miss the complicated reality that is resistant to any easy, linear narrative. Patent protections that some contend have stimulated agricultural innovation and progress emerge on closer inspection as a hotly contested arena. The remarkable scientific breakthroughs that let agribusiness firms create genetically modified seeds have arguably taken away rights from farmers around the world. Coming to grips with the associated trade-offs is not easily done, particularly on a global scale where norms around the importance of agriculture, not to mention beliefs about individual ownership of property, can be wildly divergent.

From one point of view, the development of intellectual property rights has been driven by technological advancements in agricultural technology. Genetic modification may have required clarification of existing intellectual property rights, but these more well-defined property rights may help to facilitate the creation of new technologies. Because of declining public investment in scientific research, agricultural innovation is done primarily by a small handful of private companies. For most of history, farmers easily and often shared seeds, because when you planted a seed it produced more. But once hybrid seeds were introduced, agricultural production dramatically changed. Because hybrids do not reproduce their superior traits in subsequent generations, farmers started to buy new seeds each year rather than seed saving. This meant that plant breeders and seed companies became more important. The move to genetically modified seeds has exacerbated this change. The strength of intellectual property law in the United States incentivized increasing privatization of the seed industry. Through the expansion of patent law to plants, it became profitable to invest in seed varieties where previously seed-saving traditions removed economic incentive. The major companies choose to pursue innovations that are certainly patentable, so as to increase the chance that these innovations will also be stupendously profitable. This incentive for agribusiness firms to use technology to innovate could be invaluable to the global food system.

As the Bolivian government has demonstrated with quinoa, intellectual property can also be used to preserve a country's identity and strengthen its self-reliance. Quinoa was domesticated for human consumption thousands of years ago in the Andean mountains. It became a popular 'super grain' starting in the 1980s because it is a rich source of dietary vitamins and minerals, relatively high in protein and low in gluten and is easily cooked, like rice. As a result of quinoa's tolerance of both cold and drought conditions its cultivation expanded to Kenya, India, North America and Europe. Though its rising popularity and prices signalled potential rising income for Andean indigenous farming communities, there were also downsides. As demand caused prices to spike, the poor communities who farmed quinoa were tempted to sell their entire crop and remove the staple grain from their own diets, in exchange for less nutritious and less expensive options.

Bolivia had shared quinoa's germplasm with researchers for decades but when Colorado State University was awarded a patent for their work on a hybrid form of quinoa, tensions flared between the United States and Bolivia. Though

Colorado State University let the patent lapse in 1998, this attempted control over quinoa spurred the Bolivians to include state ownership of genetic resources in the country's 2009 constitution. As a result of Bolivia's genetic-conservation programme, the government nationalized its quinoa gene bank in the hope that it would prove itself a more democratic custodian of the resources than private, and potentially foreign-funded, organizations. By asserting its food sovereignty while symbolically decolonizing the country, Bolivia hopes to lower the risk of biopiracy and forestall the private patenting of indigenous Andean genetic material. Bolivia still wants to sell it, however. In the United Nations' press release that declared 2013 the International Year of Quinoa, the Bolivian president Evo Morales hailed it as an 'ancestral gift from the Andes to the world'.[36]

Though there are ways in which intellectual property has been used to benefit scientific advancement and the interests of nation-states, patents also have some negative consequences. This is not to say that patents are intrinsically bad, but when patents for seeds are so complex that even the scientists who work on them cannot be sure whose patents they have used, this presents a problem. Only experts on intellectual property can navigate this maze properly, searching for legal treasure while avoiding legal danger.[37] The cost of this expertise effectively prices out smaller innovators, smaller firms and the least developed and most heavily indebted poor countries. Rather than helping people innovate, the system primarily helps those who are already rich to maintain their privilege and exclude competition. Though genetic modification may have been a tremendous scientific advancement, in the thirty years of its implementation there has been a lot of herbicide resistance and pest resistance in all kinds of combinations, and little else. As businesses with existing, profitable patents, there is little incentive for the firms to pursue truly creative thinking on agricultural problems as long as there is easier money to make by tinkering with proven moneymaking products and projects. Given this perspective, the problems with genetically modified seeds have at least as much to do with the business model as they do with the science.

From this other point of view, property rights that had been associated with a farmer's control and exchange of seeds have been ceded to the multinational companies that have patented the control and exchange of biological traits expressed in seeds.[38] It is not correct to assume that Boyer and Cohen sought this outcome when they first demonstrated the potential impact of recombinant DNA modification. But subsequent legal and political decisions have created economic winners and losers. In the case of agricultural biotechnology, at its best, intellectual property reinforced the status quo; at its worst, it created a new class of elite, powerful business interests. The middle-ground position is that rather than promoting innovation and change, intellectual property rights actually strengthened the dominant position of several large agribusiness firms and weakened food sovereignty efforts.

The patent controversy surrounding *Azadirachta indica*, also known as Indian lilac or neem trees, a culturally significant resource in India, is illustrative. In 1995 the European Patent Office awarded a United States chemical

conglomerate, W. R. Grace & Co., a patent for a neem-based
bio-pesticide designed to suppress insect feeding behaviour.
The Indian government challenged the patent ruling on the
basis that the neem tree and neem oil had been used for
their insect-repellent, cosmetic, medicinal and anti-fungal
properties in India for more than 2,000 years. After ten years
of legal manoeuvring, finally, in March 2005, the European
Patent Office revoked the neem patent. At stake was a question of what constituted 'prior existing knowledge' for patent
applications. Before this series of appeals, prior existing
knowledge was only recognized if it had been published in
a journal, rather than passed down through tradition. This
case helped establish that traditional knowledge systems can
be a means of establishing 'prior art' and thus can be used
to destroy claims of novelty and inventiveness. For many
observers this case was a simple act of thwarted bio-piracy,
where corporations attempt to use the patent system to
profit from indigenous products and knowledge. The fear
is that large agribusiness firms can use patents to plunder
the resources of the developing world. Jack Kloppenburg,
Professor Emeritus of Community and Environmental
Sociology at the University of Wisconsin–Madison, wrote
that 'vast tracts of the genescape and its products – DNA
sequences, exons, introns, individual mutations, expressed
sequence tags, single nucleotide polymorphisms, proteins,
protein folds, parts of plants, whole organisms, whole
classes of organism – are being appropriated via patents.'[39]
As Vandana Shiva sees it, this patent system functions as
an 'instrument of colonization' that is designed to keep
developing countries poor.[40] You do not have to believe that

intellectual property was designed to steal from the developing world to understand that this has been the consequence at times. Even if patent laws stimulate innovation, something that economists and others still debate, their unintended consequences merit attention.[41]

When thinking about the challenges in our food system, including new challenges brought about by the introduction of biotechnology, few people focus on intellectual property. This lack of focus is problematic. Currently, agribusiness firms appear committed to ratcheting up intellectual property safeguards and securing a variety of patents, while reformers call for abolishing the entire patent system altogether. Both ideas are impractical, guaranteeing that both sides will continue to disagree about even the most basic terms. Instead, committing to retaining a public balance in property rights would be a welcome change.

The concentration of research in biotechnology in a handful of multinational agribusiness firms, coupled with the development of international patenting, has fundamentally reshaped the global food market. It is clear that science, technology and industrialization have profoundly changed agriculture. What is unclear is whether seed patents are really beneficial for the public, or just for the giant corporations such as Monsanto that own and enforce them. If anything, it seems that agribusiness firms are guilty of overreaching to the detriment of innovation. Finding the appropriate balance will be a fundamental challenge for the foreseeable future. Some firms try to help tilt the scale towards public good by donating patents – a commendable act in itself; every agribusiness should be so charitable sometimes. But justice aims to create a social order in which, if agribusiness chooses not to be charitable, people are not left wanting. As the American journalist and political commentator Bill Moyers said, 'charity depends on the vicissitudes of whim and personal wealth; justice depends on commitment instead of circumstance'.[42]

3
SCARY INFORMATION?
LABELLING AND TRACEABILITY

The standard debate about labelling genetically modified food has been 'Should we or shouldn't we?' More than sixty countries, from Australia to Vietnam, representing 40 per cent of the world's population, now have some form of labelling policy in place for genetically modified food.[1] Where these laws are in place, such as in the European Union, directives indicate precise label wording and placement, leaving little or no room for interpretation.[2] So, it might seem simple to label GM food, prompting the obvious question: what would need to be done if all consumers wanted genetically modified food labelled as such? As with most of the debates surrounding GM food, seemingly simple questions lead to complicated discussions. Pithy binaries make for great sloganeering but the potential implications of a decision for or against labelling are more nuanced. Reality is not nearly as poetic and neat as political opponents would have us believe.

Many anti-GMO activists and consumers argue and insist that mandatory labelling of GM food is part and parcel of their right to know what they are eating or drinking and their right to choose what foods to consume based on that knowledge. The basic idea behind this type of compulsory labelling is to provide product information and consumer choice. Opponents note that this information comes at a cost, which depends upon several labelling characteristics, such as threshold levels, the capacity of the industry to comply with requirements and a government's capacity to enforce and implement the labelling rules.[3] Any form of labelling, whether for GM or non-GM products, will lead to additional monetary costs. Whether these costs are made visible at supermarkets and in consumer goods is not clear, however.

Food companies routinely change the wording on their packaging either once or several times a year, but the cost of such changes is minuscule. Companies can always just add an additional label or design new, updated packaging at a very small financial cost. Producers also often reformulate their products without increasing retail prices. So if companies choose to reformulate products so that they do not contain genetically modified ingredients, the cost of the food item does not necessarily need to change. For example, the original Cheerios breakfast cereal is now produced by General Mills without GM ingredients, but the price of Cheerios did not increase after the reformulation. In reality, the prices that consumers pay at supermarkets are based on several factors beyond the item price, like shopper demographics, brand competition and store characteristics. Therefore, although consumers might expect slightly higher prices, producers and other firms in the commodity chain could reasonably be expected to absorb some of these costs.

Some might argue that it is simply a matter of working out the scientific facts surrounding GM foods. Are they safe, or not? But even if there was agreement about the relevant science that could inform the process, it would not be determinative. Food labelling is less about science and more about consumers' concerns. The issues around labelling demonstrate how narrow technical issues, at one time only of interest to scientists, policy wonks and government regulators, have become imbued with differing social, cultural and political values. Moreover, any international standard will clearly create economic winners and losers and further apportion political power. The vast majority of countries with labelling policies are more economically developed, whereas only a few developing countries have introduced labelling laws. The policies themselves also differ widely in their nature, scope, coverage, exceptions and degree of implementation. Consequently, the effects of these policies also vary significantly.

One way to accomplish harmonization is to make use of international standards. For example, the United Nations Food and Agriculture Organization (FAO) and the World Health Organization (WHO) prepared joint expert consultations about genetically modified foods in 1990 and 1996. The 1990 consultation deemed modern genetic modification to be part of a continuum of modern breeding techniques, and that GM foods were not inherently less safe than non-GM foods.[4] The 1996 consultation recommended that substantial equivalence, an assessment of a novel food to demonstrate that it is as safe as its traditional counterpart, should be an important component in the safety assessment of foods and food ingredients derived from genetically modified plants intended for human consumption.[5] Determining a GM food to be substantially equivalent to another food does not mean that the products being compared are identical in every way; it means that they are identical only in the safety assessment's comparison traits. The intent of these international scientific and technical consultations is to provide evidence that will allow governments around the world to reach some agreements that will help facilitate international trade.

The Codex, short for the Codex Alimentarius Commission, was established in the 1960s as a joint inter-governmental body to facilitate the trade of food by setting international standards. Started by the FAO in 1961, joined by the WHO in 1962, the Codex held its first meeting in 1963. All members and associate members of the FAO or WHO who are interested in international food standards are allowed to fully participate as members in the Commission. The Codex may grant 'observer' status to international non-governmental organizations that have formal relationships with either the FAO or WHO; alternatively, international non-governmental agencies may apply for this status if they wish to participate in the work of the Codex on a regular basis. Observers can attend Codex meetings, receive all working documents and discussion papers, participate in discussions and submit written statements on matters before the Codex. The governments that elect to become members of the Codex are the only ones permitted to vote at official proceedings. Each member nation has only one vote, regardless of the size of its population or industry. In total, there are currently 185 member countries, covering 99 per cent of the world's population, and 233 observers.[6]

The Codex attempts to reach consensus decisions to protect consumer health and to ensure fair trade practices involving food, but one fundamental concern raised by critics of the Codex is the drastic imbalance between the prominent number of 'observers' sponsored by industry and the relative paucity of groups that act solely in the public interest. Industry-sponsored groups make up an overwhelming preponderance of observers, in part because industry groups have the financial resources to send their delegations to the international meetings.

The Codex's guidance documents are voluntary in nature, so countries can choose whether or not to adopt them as domestic law. However, the Codex standards are important in the context of international trade. The WTO designates Codex food standards as authoritative under the GATT, so their policies have major international trade consequences. Moreover, the Food Standards Programme involves the determination of priorities and provides guidance for the preparation and finalization of standards that are referred to in the WTO Sanitary and Phytosanitary (SPS) Agreement, and which are published either as regional or worldwide standards. This means that member countries are encouraged to incorporate adopted Codex standards into relevant domestic rules and legislation. Member countries can also choose to unilaterally impose more stringent food safety regulations, if they deem them necessary to ensure domestic consumer protection and provided that the different standards are scientifically justifiable and otherwise consistent with WTO SPS rules.

In 1993 the standing Codex Committee on Food Labelling (CCFL) agreed to work on the issue of GM food. The CCFL debated for more than twenty years to resolve the diversity of opinions among its member countries.[7] Following nearly two decades of discussion on the topic of biotech labelling, the Committee was only able to reach consensus on a document that compiled various Codex food-labelling texts applicable to biotech foods. To be more precise, by 2011, rather than completely abandon the project, the Committee arrived at a consensus regarding a title, purpose and relevant considerations for a document that would compile Codex texts relevant to labelling of foods derived from modern biotechnology.[8] In the end, the document gives no new guidance for countries that want to implement labelling.

In 2000 the Codex Alimentarius Commission also established an ad hoc intergovernmental task force in order to develop standards, guidelines or recommendations, as appropriate, for foods derived from biotechnology or traits introduced into foods by biotechnology. The task force was re-established in 2004 and then dissolved in 2008, after completing its work one year ahead of schedule. Though this process took several years, it reached some consensus relatively quickly in comparison to the other committee's attempt to find agreement on genetically modified food.

As a result of the decades of debate and international meetings, the Codex Alimentarius Commission simply re-affirmed that Codex texts apply to all foods, including those derived through modern techniques of genetic modification.[9] The Codex does not privilege voluntary or mandatory labelling regulations. It also does not weigh in on the consistency of any particular labelling framework. In practice this means that nation-states can choose to enact either a voluntary or a mandatory genetically modified food labelling regime, without fear of a lawsuit brought before the World Trade Organization. While the agreement falls short of the anti-GM movement's desire for mandatory GM food labelling, it does mark a significant milestone in its affirmation of the right of nations to label GM food. In other words, the Commission attempted to reconcile the competing social, cultural and political values and decided that harmonization was not possible. Instead, the continued patchwork set of national and international standards will continue, in the eyes of international law, to live an uneasy coexistence.

For example, countries that produce and export genetically modified crops must maintain two distinct supply chains if they wish to receive a premium for non-GM crops. Keeping these supply chains for GM and non-GM crops separate and pristine is not easily accomplished. Following a number of precautionary steps along the supply chain is necessary to prevent accidental commingling. Crops should be grown with buffers far enough away from genetically engineered seeds to prevent cross-contamination; harvesting equipment needs to be cleaned when switching between GM crops and non-GM crops – alternatively, harvesters could use separate equipment for each type – and all transportation, processing and manufacturing facilities would need to follow similarly thorough cleaning procedures; and finally, local and national regulators would need to declare clearly agreed-upon, understood and achievable standards for testing to ensure compliance. The local standards then have to be reconciled with international trade standards and practices. Beyond these practical and logistical steps, issues of cost, truth in advertising and fairness also need to be considered.

Throughout the process, advocates and opponents attempt to sway the opinions of regulators and consumers.

Current international regulatory regimes, which range from mandatory process-based approaches to labelling, to voluntary approaches to labelling, and everything in between, are a product of these conflicting webs of influence. Although a large number of countries import agricultural commodities, relatively few export them, and even fewer export the genetically modified varieties of such commodities. There are various approaches to labelling even among countries with mandatory labelling provisions.[10] One major difference among countries with mandatory labelling is the regulation targets – they either indicate the presence of GM content in the finished product (such as in Australia, New Zealand and Japan) or on GM technology as a production process (such as in the EU, Brazil and China). For example, in the EU any product 'produced from' a GMO must be labelled, even if there is no detectable DNA or protein from the GMO in the final product, as is the case with soybean oil.[11] However, EU regulations do not require products derived through the use of genetically engineered yeast and bacteria – such as beer, wine and cheese – to be labelled, because such products are not 'produced from' GMOs.[12] On the other hand, Australia, which also has a mandatory labelling system, does not require labelling in the absence of detectable DNA or protein.[13] As a result, soybean oil, which contains no detectable DNA or protein, must be labelled in the EU, while no label is required in Australia. In the end, global food traders are faced with the dilemma of increasingly common GM commodity crops, a complex patchwork of regulatory hurdles and unresolved consumer sentiments regarding GM foods.

In total, more than three billion people live in countries with laws in place, planned or proposed, to label or ban genetically engineered food. Mandatory labelling in European nations and several other countries, including Japan, South Korea, China, Australia and New Zealand, is drawing greater scrutiny to this issue. This creates an increasing need for product traceability – which enables GM commodities and their products to be followed throughout the production chain – and analytical testing. Genetically modified foods can be detected by analytical tests since they contain unique novel proteins and/or DNA sequences that are not inherent in their conventional counterparts. Traceability facilitates the withdrawal of products when unforeseen adverse effects on human and animal health or the environment are detected, as well as aiding their continuous monitoring in order to examine potential long-term effects on the ecosystem.

Traceability and testing for the presence of GM ingredients are foundations to labelling GM food. Some propose that a mandatory labelling law could benefit agricultural exports, while some have suggested that even voluntary labelling could increase industry credibility and consumer acceptance. Proponents of labelling note that it is accepted practice to label food products in terms of additives and colourants – even though these do not pose any health risk – as well as religious (that is, halal and kosher) preference or lifestyle choice (whether a person is vegan or vegetarian) without any consideration of cost. Furthermore, proponents argue that voluntary (and not mandatory) GM labelling gives discerning consumers a choice without prejudicing non-discerning consumers in terms of cost.

The potential benefits of labelling do come at a cost, however. The accurate segregation of GM foods through all phases of production (planting, harvesting, processing and distribution) would add to food production costs and potentially compromise economies of scale. These added production costs could mean reduced profits to plant breeders, farmers, food processors, grocers and others in the distribution pathway, leading, moreover, to price increases for consumers. A problem to consider with regard to the application of voluntary GM labelling is that, throughout the world, it is not currently regulated and may result in consumers being misled. Consumers may believe that 'GMO-free' and 'non-GM' labels indicate an absence of genetic modification, despite the fact that no regulation-based definitions exist for these terms. The absence of specific definitions for voluntary GM labelling in some countries is exacerbated by the use of these terms in a mandatory context in other countries.

Even deciding upon the allowable degree of purity for non-GM products is a contested activity. For many products it would be unreasonable to demand a complete absence of GM ingredients. Therefore, countries grant some degree of tolerance to account for the adventitious presence, the accidental or unintentional appearance of genetically modified ingredients, in a product. But the degree of acceptable tolerated traces of GM ingredients varies. For example, the EU applies only a 0.9 per cent threshold for 'non-GM' food; by comparison, the threshold is 5 per cent in Japan and 3 per cent in South Korea. The specific units used for the percentage calculation may induce disputes between countries and among national stakeholders of a supply chain. Regardless of the threshold, any form of traceability and labelling requires a complex infrastructure of monitoring, detection and verification. In response to the complexity and costs of stringent labelling and traceability standards, and related efforts to realize the coexistence of GM and conventional agriculture, some countries have instituted moratoria or bans.

The idea that food should be traceable, from source to plate and back again, is only partly related to genetically modified food. The bigger impetus comes from food safety directives related to a number of food safety scares, ranging from mad-cow disease and the related Creutzfeldt–Jakob syndrome scares in the United Kingdom, to dairy product contamination in China and the H5N1 avian influenza (HPAI H5N1) circulating in Asia, Europe and Africa. The lack of traceability is a problem not only for consumers, but for firms in the supply chain. Traceability cuts across a wide range of food issues relevant to trade, within the context of risk management. Whether traceability should be used for consumer-based information, like labelling, is more controversial.

Let us consider recent attempts in the United States to label genetically modified food. Although recent ballot initiatives in California and Washington narrowly failed, the efforts to label foods containing genetically modified ingredients have taken on added importance. The states of Vermont, Connecticut and Maine have recently passed labelling legislation, and roughly half of all state legislatures are considering labelling requirements for GM food. Labelling GM food holds personal appeal for consumers because it gives people the illusion of being in control of their own food choices. There are dangers, however. If labelling proponents are successful, they may unintentionally contribute to the very problems they wish to counteract. Even a national labelling law for GM food might have relatively modest impacts on the American food system given the massive shift to genetically modified commodity crops in recent decades.

Even if one agrees with the ideal of product transparency, Andrew Szasz's theory of inverted quarantine – explained in *Shopping Our Way to Safety* as when a consumer believes they have shielded themselves from a perceived threat (that is, GM food), therefore producing a sense of security – shows how complicated the issue is, and how incorrect it may be to assume that labelling GM food represents a victory over the modern, industrial system of food production.[14] For example, in a poll, a sizeable majority of Americans declared their desire to see GM food labelled as such, but a considerable number also approved of mandatory labelling for food containing DNA.[15] So it is not clear exactly what consumers want to see on their labels. It is possible that their responses could stem from a general desire to obtain more information about the foods they eat. Even though almost all Americans, when asked directly, report that they believe GM foods should be labelled as such, most do not mention genetic modification when asked more generally what information they would like to see on food packaging.[16] In this sense, consumers' commitment to GM food labelling is shallow – they only wish for it, or consider it important, when prompted to do so. In other words, this issue lacks salience. Similar, albeit smaller, numbers of consumers want to know where the food product was grown or processed.[17] While many are concerned about the lack of labels on GM food, most consider other issues such as hormone, pesticide and antibiotic use more pressing.

How do you explain the gap between people who simply want GM food labelled, and those who care enough about it to make a commitment to a system of food production that does not rely on genetic modification? Mass belief in personal choice and personal ability to avoid an undesirable product is part of the answer. With labelling, people may come to believe that they have successfully insulated them-selves from the problems of agriculture. As Szasz deftly argues, consumers continue to care about the quality of their food, but if it were fully labelled, they might lack any real motivation to actively do anything more.

If genetically modified food were labelled, some people might avoid such food in order to 'protect' themselves, and will thereby be less likely to feel the urge to voice support for the kind of regulatory controls required to fully address structural issues in the food system. As anyone who has looked at the latest farm bill can tell you, U.S. federal food regulations and politics are fraught with food industry interests and influences. Without broad support, regulators have little incentive to go against the powerful influence of agribusiness. The mass flight of consumers into inverted quarantine, brought about by labelling, might decrease the chance or defer the possibility that something substantive is done about problems in the food system. Robbed of a potential mass constituency of highly motivated supporters, political motivation will languish. In this sense, labelling might impede the development of a significant mass of consumers who are committed to critical thinking about the food system.

Although individual consumers might try to change their shopping habits to avoid foods labelled as containing genetically modified ingredients, they are not able to do so easily. If a more just or sustainable food system is the eventual goal, labelling laws can only go so far. Labels will not fundamentally alter the structure of food production. And, if we believe in an inverted quarantine, labelling laws might actually hurt the movement. But this is just a theory. There is a vigorous social science debate about this phenomenon, what some call ethical consumption, political consumption or green consumption. Some argue that this type of consumption choice based on labelled foods is actually detrimental, because it leads people to think that this purchasing behaviour can solve problems, and it leads them to be less likely to join in collective solutions to the problems of environmental destruction, labour exploitation, poverty and stunted development in the global South. Others, however, observe a strong correlation between people engaging in this kind of purchasing and being socially and politically involved in the collective effort to solve such global problems.[18] This relationship between consumption and action has a strong link, whether people consciously consume and then become politically active or whether they become politically active and then begin consciously consuming. The idea that becoming a conscious consumer will make you less likely to engage in collective action is not empirically supported.

Changing marketplace behaviour can be an important component in a broad-based campaign for labelling GM food. As can be seen in the two decades of international debates surrounding GM food, it can be nearly impossible to get nation-states to act given the competing agendas at play. This means that people have turned to the marketplace instead, in part because it is an arena where it seems as if you can achieve some results, at least in the short term. Market behaviour will ultimately be insufficient. Rather than changing individual consumer behaviour at the point of purchase, truly meaningful change must start long before products hit the shelves.

Too often, the international food system relies on a Hobbesian worldview: the race for lower prices and individual choice to the detriment of broader societal benefits. It might be better, for individuals and society as a whole, for the public to shun the inverted quarantine made possible by potential GM food labelling – a short-term fix. Instead, the public may want to concentrate on substantial food system reforms that address the issues surrounding the broad availability of diverse and nutritionally adequate sources of food; intellectual property; national sovereignty and colonialism; consolidation in the agricultural chain of production; and the regulation and management of environmental hazards.

Even when all sides champion the principle of protecting individual interests, agreeing on how to achieve that goal is fraught with discrepancy and conjecture. After decades of debate, even meagre agreements are illusory. It could be that the search for a best possible set of national or international standards allowing fair trade, consumer choice and innovation is a worthy and commendable endeavour, but these considerations obscure the deleterious structural conditions of the global food trade. Given international interdependency, we may find, for example, that we cannot deal with issues of GM food without addressing inequality; poverty and other types of inequality strongly influence people's power to control the benefits and risks they face in life. Seed companies may believe they can solve agricultural problems like pest and disease resistance by creating technocratic solutions, but these may only drive problems across national borders; that is, if rich states or nations impose strict controls and regulations over planting, those crops are relocated to somewhere with weaker regulations. This does not solve agricultural problems as much as simply shifting the potential risks onto another group of people. If we do not attend to agricultural production practices at a base level, then menu choice will continue to look quite different for the affluent compared to the impoverished.

Moreover, even if genetically modified food were to be labelled as such, it is unclear if that label would be consequential. The meanings of labels often change dramatically as the certification system behind them evolves.[19] As Julie Guthman noted in her book *Agrarian Dreams: The Paradox of Organic Farming in California*, large corporate farms have systematically diluted the meaning of 'organic'.[20] Similarly, the inclusion of large corporate buyers of fair trade products has grown the market significantly while funnelling economic benefits to producers. However, despite the fair trade movement's intention to challenge the dominance of large transnational buyers, it penalizes small buyers disproportion-ately.[21] In both cases, as corporate agribusiness firms continue to enter the organic and fair trade sectors of the market, they tend to seek regulatory changes and weakened standards that favour their existing business models. In doing so, they co-opt alternative agriculture and food movements, de-emphasizing the most transformative elements of the original standards.[22]

This debate is similar in many ways to debates over the use of the word 'natural' on food packaging. There is no official or legislative definition of what products can claim to be 'natural'. Trade groups, lawyers and government agencies have proposed a variety of ways to resolve the confusion, ranging from defining the term to suing over it and ignoring it, respectively. Some consumer advocacy organizations simply call for a complete ban on the use of the term in labelling. The reality of our modern food system is that it is difficult to define where 'natural' ends and where corporate food science begins.

Orange juice provides an eye-opening example. To make freshly squeezed orange juice, you need to cut a few oranges in half and squeeze the juice into a glass; you can use a citrus reamer or hand juicer and strainer if you prefer. Each glass may taste a bit different, depending on the variety of oranges, the time of year and the quality of the fruit. Though it is not terribly time-consuming to make yourself, most consumers prefer to buy orange juice. Producing juice on an industrial scale means performing a slightly different set of tasks than you would at home. After the oranges are washed, squeezed and pasteurized – heating the juice for a short time to eliminate potentially harmful bacteria – the firms store the juice in million-gallon de-aerated aseptic holding tanks.[23] Oxygen is then removed from the tanks so that the juice can be preserved and dispensed for up to a year without spoiling. Before packaging and shipping, 'flavour packs' made from the chemicals that make up orange essence and oil are added to the juice to compensate for the loss of taste and aroma during the heating and storage process. Food scientists break down orange essence and oils into their constituent chemicals and then reassemble the individual chemicals in configurations that ensure a distinct flavour profile for each brand of orange juice, provide consistency from month to month and match the preferred taste of consumers in different countries.[24] This thoroughly modern and industrially produced orange juice is marketed as 'pure', 'simple' and made from '100% juice'. But is it – given a global system of agriculture that demands the year-round availability of foodstuffs – as natural as possible? Given the amount of food science involved in 'natural' products, how do we distinguish truly natural from nominally processed?

There are also logistical and practical questions associated with labelling, given the global reach of the agricultural seed firms that are involved. The power wielded by these firms is evident in many ways, from their ability to buy or eliminate competitors to their ability to mobilize nation-state power, mostly aligned with the United States' interests. Some might even argue that the growth of agribusiness power, and its entwinement with, and influence on, the u.s. government, has led to a situation in which consumer interests and well-being cannot be adequately ensured. As such, firms are unlikely to be forced to embrace global, mandatory labels for GM foods if they perceive this is against their economic interests. Labels also raise the possibility of negative unintended consequences for those who crave them most. A voluntary or weak standard for labelling might abate general support for a movement to eliminate GM food. It is reasonable to expect that an industry would resist obligatory labels that could cast its products in a negative light. But corporate interests can change, particularly if they can market those changes to consumers as positive developments, and, as demonstrated by European activists, countries can institute labelling rules and other restrictions, even in the face of considerable industry resistance.[25]

Importantly, the concentrated ownership of organic food and beverage brands by multinational corporations such as Kraft, PepsiCo, Unilever and Nestlé may act as a countervailing force to the desires of seed and chemical firms like DuPont, Monsanto, Syngenta and Groupe Limagrain. Food processors and marketers believe that demonstrating concern for consumers can burnish their reputations with investors.[26] So as social movements that demand the labelling of genetically modified food drag on over time, firms can create innovative processes and ingredient sourcing internally, redesigning and reformulating their products concurrently with the design of regulations. Food marketers can segment their product offerings so that consumers might be willing to pay more to get or avoid genetically modified ingredients. This means that food companies could view labelling as a profit-making opportunity, not a threat, and devise marketing strategies that work with labelling policies.

Where does this leave us? Given the status quo of consumer confusion and expressed desire for transparency, even a voluntary labelling effort might help come up with a meaningful definition that does not undermine other efforts. This would be particularly consequential if the labelling guidelines are the result of a consensus-based process in which all stakeholder interests are considered. A voluntary, strong definition adopted by several sectors of the food industry could provide a reasonable minimum starting point for a broader-based regulatory approach. This is not a panacea; traditional food politics dictate that even this would be a less than fair outcome for consumers. Therefore, litigation must remain a viable, if imperfect, way to hold agribusiness firms to some degree of transparency.

This is not an ideal account for either side of the debate. The true impacts of food policy are rarely satisfying to every-one. The reality is a messy balancing act of political, cultural, economic, ethical and scientific needs. It is also often banal, and hidden in the wearying language of trade agreements and international policy accords rather than expressed in compelling personal narratives. Displaying information on product labels might impact individual consumption; it will also likely govern production in anticipation of individuals altering their consumption. If global regulators use their capacity to demand transparency for the purposes of food safety, they may also wind up governing industrial production and trade policies.

4
SCIENTIFIC FALLIBILITY: CONTESTED INTERESTS AND SYMBOLIC BATTLES

Let us try to place the evidence surrounding genetically modified food in a broader context. Why is it that the majority of the scientific community accepts GM food but the public remains sceptical? There are those who do not see the scientific understanding of GM food and opposition to its adoption as contradictory. Science is filled with contested interests and symbolic battles that help define what is considered acceptable and what is considered shocking. One of the roots of the controversies surrounding genetically modified food is that both parties claim that they have access to some kind of incontrovertible truth. One group's certainty springs from ethical, social and cultural insights; the other's derives from a no less dogmatic scientific perspective. The debate between the two endures, and frequently degenerates into a clash of unexamined preconceptions and worldviews that are rarely reconciled.

Science, like the human beings who conduct it, is fallible. No matter how careful the observations or reasoning, all scientific theories are incomplete; no matter the scientific domain, there are always non-trivial scientific questions still to be answered. Moreover, scientific measurements are always accompanied by some degree of error. As such, scientific 'truths' are conditional and not absolute. Even easily observed, easily measured and unambiguous data does not convey a single, unequivocal message. Consider the idiomatic question 'is the glass half empty or half full?' Optimists and pessimists can use the same level of water in a glass and yet come to opposite conclusions. In other words, data never speak for themselves. This is the beauty of the scientific method. Theories, even those that are commonly treated as facts, remain active areas of scientific inquiry, with many fascinating questions remaining. No single study produces a definitive answer. New scientific evidence can help researchers to refine, support or not support their hypotheses; the evidence itself is produced over the long term and often involves false starts, dead ends, minor successes and a slow, grasping, incremental production of knowledge. In this sense scientists are never able to prove the absolute truth of their position. That is not to say that science is a form of 'truthiness', where scientists only talk about something that seems like the truth they want to exist. Instead, this continual re-examination of evidence is exactly what makes a theory scientific. If a statement cannot be falsified it simply is not part of a scientific inquiry.

Scientists, and those that follow the scientific method, use evidence to construct testable explanations and predictions. Because observations and explanations build upon each other, science is thus a cumulative activity for constructing knowledge. Through repeated observations and continued experimentation, scientists describe the world more accurately and comprehensively. One generation's observations and explanations suggest new questions to be tested, and provide new extensions to our understanding. Each subsequent generation of scientists work to correct, refine and expand upon the work completed by their predecessors. Isaac Newton's famous statement that 'if I have seen further it is by standing on the shoulders of giants' illustrates the idea that all knowledge and scientific discovery is cumulative; the sophistication and scope of scientific explanations improve over time.

In his 1912 Nobel Prize lecture, the French chemist Paul Sabatier said that theories 'are only the plough which the ploughman uses to draw his furrow and which he has every right to discard for another one, of improved design, after the harvest'.[1] New scientific methods and tools sometimes reveal the inadequacy of a discovery; similarly, new ideas sometimes reveal the insufficiency of prior explanations. Even more, some scientific ideas that were once generally accepted are now understood to be completely wrong or to only apply within a limited domain. For example, Aristotle and Ptolemy famously theorized that the Earth was the centre of the universe. More than 1,500 years later Copernicus proposed the opposite, the heliocentric theory of the planets orbiting the sun. Today, however, modern astrophysicists treat the sun as neither a stationary body nor the centre of the universe; instead, the sun rotates around the centre of our galaxy, and in turn our galaxy is in constant movement within the cosmic background.

As with any active area of science, figuring out which questions are worthy of ruthless, critical inquiry is at the crux of the matter. Though simple in premise, the process can be difficult. Given that science depends on empirical evidence and testable explanations that accumulate over time, it should not be surprising that many fascinating scientific questions about genetically modified food remain unanswered. It is, after all, a relatively new technology. Moreover, because food is so personal and because the stakes are high, the GMO debate can seem too acrimonious to allow an honest discussion. Many people also have conflicted feelings about food as a topic of focused scholarly attention: 'Food is the first of the essentials of life, our biggest industry, our greatest export, our most frequently indulged pleasure, and also the object of considerable concern and dread.'[2] The passion expressed on all sides of the debate scares some people away. Researchers who make positive remarks about biotechnology are accused of being corporate shills with vested interests in the success of agribusiness firms. Similarly, those who question the human health effects or environmental safety of GM crops are subject to professional ridicule and scorn. Attempting to treat either side of the debate with respect often angers both. Nevertheless, dispute and divergent opinions are part and parcel of scientific and societal progress. As Mahatma Gandhi said, 'honest disagreement is often a good sign of progress.'

Surveys show that many ordinary members of the public continue to be sceptical about genetically modified food. These same surveys have also shown, and continue to show, that the public's scientific understanding, particularly of genetics, is incomplete, incorrect or in doubt. Many have never even heard of traditional forms of cross-breeding and are not aware that they eat traditionally cross-bred fruits and vegetables as a matter of routine. It makes one wonder how people would respond to Jean Anthelme Brillat-Savarin's famous aphorism: 'tell me what you eat, and I will tell you what you are.'

The modern food world is filled with a vast array of uncertainties for the general public. Is there really a need for GM food? Is it needed to help feed the world? If not, how does it help? Do the companies who create these seeds believe that profits are more important than safety? Did scientists create genetic modification techniques to satisfy a public demand or was it simply because they were able to do it? Can we predict what will happen with GM crops or is nature so complex that it is impossible to anticipate? Are serious accidents bound to happen? Should we regulate them more, or less, strictly? Is the technology positive or negative overall? Is it positive or negative overall for me? For society? Is there a particular or limited application of the technology that would be beneficial? For me? For society? For the environment? Do the people who are involved with this technology share my values? Can I trust them?

Given the public's lack of scientific knowledge and lack of familiarity with the global food chain, it is tempting to conclude that the public is incapable of reaching informed conclusions about any of these questions. But that would be wrong. The public's social, cultural and ethical values are unlikely to match those of scientific experts, but that does not mean that their beliefs are any less valid. The general public is likely to reach different conclusions from whatever evidence is available. Scientists often communicate using numbers; non-scientists, and most journalists who convey information to the public, communicate using words, stories and visual signs. These different forms of communication often present a barrier to understanding. This barrier also exists between experts in different fields. Agricultural economists are likely to be interested in different questions than food scientists, for example. Both fields are interested in improving food production and processing, but they have radically different approaches to the problems. In reality, the knowledge that agricultural economists, food scientists and other scholars add to the debate is important, but educating people about the scientific details does not necessarily lead to greater comprehension of the big picture or the ability to make informed decisions.

Seeing the big picture is not simply a matter of observing all the details. Often, to grasp the details you need the context of the big picture. So, it does not seem that it would be really helpful to try and move people from a position of using intuition or initial reaction to GM food to the more considered form of evidence-based decision-making advocated by many scientists. That simply does not reflect the entirety of how people make real-world choices – choices which are influenced by many things: the impact of new information and evidence, as well as corporate advertise-ments and enticements; the judgements or decisions made by others they respect or know; and the sense of social desir-ability. Fad diets, food trends and friends' recommendations influence us tremendously, bringing culinary obsessions including the cronut, kale and the Paleo diet. When potential risks like those of GM food are involved, people also weigh a number of other factors including beliefs about nature, the fallibility of science, trust in the organizations promoting the products and cultural norms, for example. Even if science and evidence were to all point to the safety and beneficial importance of GM food, it still might not be convincing enough to sway the public.

Given its contested status, it would seem reasonable that a large-scale adverse event involving GM food could likely serve as a pivot point that would quickly catalyse public opinion and decision-making against the technology. From the other side of the debate, it is hard to imagine a positive development that would be likely to turn public opinion and decision-making in favour of the technology. In the big picture, the public remain so sceptical of this technology and the groups who promote it that they are willing to believe that a 'large fast-food company used chickens so altered by genetic modification that they can't be called "chicken" any-more'.[3] This story is, of course, false. But the fact that the public are unsure of its veracity or are outright willing to believe it to be true shows both the public's low level of understanding of their food system and their lack of trust in it.

Whether the public has a legitimate reason to mistrust the food system is almost beside the point. In a sense, it is true that terrible things are unlikely to happen to our food system. For all of the food safety scares and even in cases of purposeful adulteration, consumers have been kept safe and the regulatory system has performed its function. Still, as the American sociologist Lee Clarke reminds us, just because there is a low probability of an event occurring does not mean we should ignore it.[4] Just think of the Fukushima Daiichi nuclear disaster. The odds that the power plant would be inundated by a massive tsunami like the one in 2011 might have been small, but considering the possibility and taking appropriate steps might have prevented the disaster that ultimately happened. Post hoc analysis indicates that protecting emergency power supplies by moving them to higher ground or by placing them in watertight bunkers, establishing watertight connections between emergency power supplies and key safety systems, and enhancing the protection of the seawater pumps would have prevented or mitigated the worst effects.[5] Planning for the worst cases like this is an entirely rational thing to do. After all, even when things go according to plan there can be unintended and unforeseen consequences.

As with most controversies, there tends to be a range of debates and reactions. Scientists have often assumed that resistance to GM food is due to a poor grasp of scientific and technological facts. This is simply not the case. There is a remarkably weak, though not entirely absent, statistical relationship between people's attitudes towards GM foods and their knowledge of the underlying science.[6] In other words, teaching people more about the science behind genetic engineering is unlikely to change opinions about genetic modification in any particular direction.

Researchers in agribusiness firms and academia have worked diligently to use objective measures in assessing the science of genetic modification, ones that their colleagues and peers will appreciate. This is appropriate. However, because genetic modification is not purely an empirical matter, we must recognize that science alone may not be enough when assessing choices in our global food system. Different value assumptions will lead to different emphases in the food system. There is an intimate interplay between science and values where the development and stewardship of the food system is concerned. There are many who believe that more efficient food production is necessary and almost always good. On closer scrutiny, our food system is conceptually more nuanced than this and includes an ethical and value-based component. Our food system is concerned not only with efficiency, but with social welfare, personal health, sustainability, environmental values, biodiversity and cultural heritage, to name but a few examples. When the focus is exclusively, or just primarily, on scientific and economic efficiency the opportunities to express these other values are thwarted. These emphases, informed by empirical insights, reflect value frameworks that influence the composition of our global food system. By including these values in our conversations we open up discussions about how far our systems of food production and consumption go towards meeting all those goals. The result would probably be a system that borrows techniques from many existing methods. That would be good news for farmers and the rest of society. It would, however, challenge powerful stakeholders who benefit from the status quo. This is why conversations about GM food are so fraught with difficulty.

Scientific visionaries and agribusiness industry spokespeople talk about reduced consumer prices, nutritionally superior grains, and crops adapted to local conditions that can turn marginal agricultural land into productive farms. The day may come when genetic engineering lives up to that lofty potential. But even after thirty years none of those benefits are clearly evident, leaving many to grouse that genetically modified foods have delivered little or nothing in the way of tangible consumer benefit.

To date, agribusiness firms have focused their efforts on commodity grains. These productive and versatile crops have responded to investments in research, breeding and promotion. As the International Service for the Acquisition of Agri-biotech Applications likes to point out, genetic modification is the fastest-adopted crop technology in recent history, growing more than a hundredfold in less than twenty years.[7] But the potential selling points for GM foods – personal health, environmental well-being and perceived contributions to the public good – that are actually available to today's consumers are unclear, greatly limited or non-existent. Any talk of special consumer benefits to offset risk, however real or imagined, is largely hypothetical because these genetically modified products are simply not on hand. Under these circumstances, it becomes easier to understand why even minor or potential risks matter.

On the one hand, some consumers may be swayed by arguments that GM crops make the food production system more efficient. Similarly, they may be impressed by arguments that GM food is the simplest and most pragmatic means of fighting hunger, disease and poverty. They might argue that countries facing these widespread problems may weigh risks differently than in more affluent locales. Deciding whether efforts to improve food security in poorer countries should focus on biotechnology or on radical shifts in farming practices is not a theoretical exercise. But, as with most real-world problems, the issues are not so simply solved. Take, for example, the basic counterpoint that the increasing use of GM crops might disproportionately harm farmers in the developing world.

Not all concerns are economic in character. Religious objections may arise from those who feel that God created everything perfectly and, therefore, people should not change, manipulate or tamper with the status quo. Environmental objections often take a similar form – the idea that it is inherently wrong to change nature. They may also point towards the risk of altering fragile and unknowable ecosystem equilibriums; there has been a long history of damage done by pesticides that agribusiness and chemical companies purported to be harmless, and animals that have been thought unimportant in an ecosystem which, after their removal, have been re-evaluated as highly important. It is essential to recognize that narrowly conceived economic factors like product pricing do not drive all decisions or even consumer decisions. If they can afford them, people will pay for things they need or want. For as long as advertising has been practised, advertisers have relied on persuading consumers to buy, or want to buy, things they have never heard of before. That large, powerful and wealthy agribusiness firms with fantastically innovative and brilliant scientists have not fully convinced consumers that GM food is a boon is testament to the fact that concerns about GM food are about more than money or simple science.

Exploring some of the major controversies surrounding the scientific understanding of GM food is illuminating. Entities on both sides of these debates attempt to use science to discredit the opposition at some risk to their own credibility, a potentially expensive proposition in terms of the spending of social capital. Under these circumstances, as with other controversial or emerging technologies, even minor risks matter to the public; but the risks of GM food are not necessarily minor. Understanding the public reactions to science and surrounding controversies raises epistemological questions as well as socio-political ones about the authority of science in the global food system.

One of the first debates began when a scientist named Steven Lindlow at the University of California, Berkeley, discovered a bacterium that caused some plants to freeze at higher temperatures than normal. Crop damage from low temperatures costs farmers more than $1.5 billion each year, so finding something that could help prevent crops from freezing in a cold snap would be an agricultural boon. By 1982 Lindlow and his team of scientists began conducting safety experiments and planning field tests to see if their genetically engineered bacteria known as ice-minus could help crops fight frost. Over the next several years of protests, litigation and safety tests, a recurring pattern of argument became evident, setting the tone for how both pro- and anti-GM groups would define acceptable risk.

Large agribusiness firms and agricultural scientists attempt to judge whether the anticipated benefit of new knowledge from their experiments justifies undertaking the risks. Although it is impossible to rule out all possible negative outcomes, in general if the risk is low it is considered reasonable to continue the scientific exploration. This sort of incremental, considered scientific risk assessment is part of normal, acceptable progress. But what are we supposed to do when the technology and science are novel?

These pioneers of genetic modification relied on technical expertise and scientific competence to make affinities between known plant-breeding programmes, along with existing herbicide- and pesticide-testing procedures and precautions, and any uncertainties that may have existed with this new technology.

Perhaps because of public pressure, regulatory agencies required an unusually high degree of caution for the testing. The scientists wore full-body protective coveralls and respirators while spraying the test fields with handheld dispensers, and regulatory agents stood on ladders to check air monitors to make sure the dispenser's contents did not spread beyond the field's boundaries. Trespassers uprooted half of the plantings before the seedlings sprouted, forcing scientists to replant. The researchers also had to hire an overnight guard to sit in a van parked nearby. Future test plots involving ice-minus were also vandalized. Though the extra precautions, such as the protective gear, security guards and air monitoring, were likely undertaken out of an abundance of caution, they created an enduring vision of necessary precaution. After all, anything that requires all that safety equipment and monitoring must be scary. Advanced Genetic Sciences, the small biotechnology company that licensed ice-minus under the name Frostban, gave opponents another reason to be concerned. The company tested Frostban on the roof of its Oakland, California, headquarters without official permission. Though they were fined by the United States Environmental Protection Agency for their actions, they validated concerns that companies involved with genetic modification were not to be trusted.

In this sense, the abundant precaution and corporate impropriety helped validate anti-GM advocates' leeriness of the new technology. Science expects potential harms to be balanced by potential benefits. But even as global firms and modern technology have provided the public with many benefits and developments, the public has also had to contend with catastrophic failures such as the 1976 chemical release at Seveso, Italy; the 1984 Union Carbide gas leak in Bhopal, India; the 1986 Chernobyl nuclear power plant explosion in Ukraine; the 1989 *Exxon Valdez* oil spill in Prince William Sound in the Gulf of Alaska, USA; and the 2011 Fukushima Daiichi nuclear meltdown in Japan. In each of the events, all involving omnipresent modern industrial technology, one basic question remains. If all scientific progress comes at some degree of risk, what level of risk is acceptable? Anti-genetic modification activists can attempt to magnify and exploit these uncertainties to advocate for strict legal and regulatory oversight, defining any risk as unacceptable.

Taking this strong precautionary stance, however, frames the debate about genetic modification in terms of scientific risks and benefits. Though the above-mentioned industrial disasters were horrific, most debates are about how to make these industries and products safer through more stringent safeguards and regulations. Instead of deeming any of those technologies unacceptable, the emphasis has been placed on how to contend with the risks by improving education as a form of disaster risk reduction. Because scientific studies have many uncertainties, scientists must extrapolate from study-specific evidence to make causal inferences and recommend protective measures. Absolute certainty is rarely an option. In any given scientific debate, there will be various published studies with inconsistent or even contradictory findings. Agribusiness firms have also learned to exploit the uncertainty around these inconsistent and contradictory findings, labelling them 'junk science'. Often challenges to the science of genetic modification have distracted attention from the social, cultural and ethical entanglements.

For example, a relatively well-known scientific debate about GM food took place in England during 1998.[8] At the centre of the controversy was Dr Árpád Pusztai, author of hundreds of research articles on food safety and a member of the Rowett Institute, one of the United Kingdom's leading food safety research labs. He sparked a controversy when he expressed doubts about the safety of GM foods on Granada TV's *World in Action*. To illustrate his concern he mentioned his ongoing research on GM versions of pesticidal proteins. The research involved feeding two sets of rats a protein (lectin). He fed one set of rats using potatoes that were genetically modified to produce more lectin; he fed the other set potatoes that had lectin added by non-GM methods. According to the findings, the rats which fed on GM potatoes suffered a number of harmful effects on growth, organ development and immune responses; the other group of rats did not suffer the same ill effects. Dr Pusztai speculated that the GM device used to carry the new gene into the potatoes might be the source of the problem.

Following his television appearance, politicians, scientists and the biotechnology industry vigorously attacked Pusztai and his research. In a letter to the Royal Society, the United

Kingdom's national academy of sciences, he wrote, 'I have suffered allegations concerning my personal honesty and motivation; those concerned preferring to attack me rather than treat my work and findings to an informed and balanced appraisal.'[9] Neither his eminence in the field nor his careful documentation and scientific defence of his statements were enough to save his career. Pusztai was suspended after 36 years' work at the Rowett Institute, and his employment contract was subsequently not renewed at the end of 1998, notwithstanding that one to two years of work remained on each of the six research programmes for which he was responsible.[10]

Following Pusztai's dismissal, he detailed his scholarship in the 'Research Letter' section of *The Lancet*, and the editors wrote a lengthy explanation of their decision to publish the findings.[11] The article carefully maintained that the data were preliminary and not generalizable, and the conclusions were weak and tentative. Many of the scientific reviewers had concerns about the design and execution of this particular research. However, as the editor noted, the debate was no longer about the merit of the research itself, but about the framing of science and dissemination of information to the public.[12] Pulled into the debate were academics, scientific journals, various media outlets, government officials, industry executives and numerous advocacy groups.

A year later, a similar controversy emerged over preliminary research conducted at another prestigious research centre, this time in the United States. In 1999 researchers from Cornell University published a letter in *Nature* stating that pollen from Bt corn, a type of genetically modified corn, had toxic effects on monarch butterfly larvae.[13] Caterpillars, the larval stage of monarch butterflies, feed on milkweed plants. Because some milkweed grows next to cornfields, lead researcher and entomologist Dr John Losey and his Cornell colleagues suggested that Bt corn pollen may drift onto milkweed and inadvertently harm the monarch larvae. Although not a full scientific paper, the research garnered a tremendous amount of media coverage and anti-biotechnology advocates adopted the monarch butterfly as a public symbol of the environmental consequences of GM food. Though there were some initial attempts to discredit the research, a second study confirmed some of the initial scientific findings.[14] In the following year, the EPA, biotechnology industry and university researchers studied the potential impact of Bt corn pollen on the monarch butterfly and related species and found that Bt posed little risk of harm to their larvae.[15]

Despite industry mounting attacks on Losey and his scholarship, Losey himself called for more study and a measured approach to the issue. Perhaps as a result, Losey did not face the same fate as Pusztai. Again, the fight was not only about the science, but about the framing of the debate. As a marker of the initial scientific debates, monarch butterflies have come to symbolize the *potential* risks of GM crops. Until then, the official risk assessment had managed to avoid considering the effect of the Bt toxin on non-target insects. In this context, the criticism about the methodological limitations of the study reinforced its message that serious consequences can follow from unintended interactions with the broader environment. This study did not prove that Bt corn kills monarch butterflies, but it raised the question of why such experiments were not performed earlier.

In 2001 Dr Ignacio Chapela from the University of California at Berkeley, along with his postdoctoral student Dr David Quist, published a paper in *Nature* contending that pollen from GM corn (maize) had spread into non-GM corn in Mexico.[16] Just how the contamination occurred remains a puzzle, especially since Mexico had a moratorium on the planting of GM crops that had been in force for three years by the time the contaminated samples were collected. Agricultural experts and proponents of GM crops maintain that corn pollen is characteristically heavy, so winds do not carry it far from cornfields. The closest region where farmers and industry had ever officially planted GM corn was 60 miles away and therefore wind-assisted contamination was impossible. Chapela suggested that contamination might have occurred due to fresh hybridization events with illegally cultivated GM crops, or as the result of 'escaped' GM genes that had persisted in traditional corn since the government-imposed moratorium. This second possibility was controversial.

In the spring of 2002 *Nature* published letters by well-known scientists who questioned the validity of Quist and Chapela's research. With criticism and pressure coming from many sides, *Nature* took an unprecedented step. For the first time in the journal's history, the editor announced that it should not have published the article in the first place, despite the original peer review, due to insufficient evidence.[17] Even more irregular was that the major finding of that paper – that GM contamination had occurred – was never in dispute.[18] Detractors directed their technical criticisms at a secondary finding suggested by the data – that the transgenic constructs were fragmenting and scattering in the maize genome. This suggestion, that the inserted transgene is capable of moving around a genome, either intact or in fragments, sparked the controversy. Many within the scientific community agree that the claim of transgene reassortment in the genome is unsupported by evidence at this point.

Whether or not parts of Quist and Chapela's study were technically flawed, they focused attention on an important concern deserving careful analysis and evaluation. Some of the controversy occurred because maize is a staple, historic crop with immense cultural significance in Mexico. Furthermore, Mexico is the world's repository of maize genetic diversity so this threat gave a vibrant, real-world example to match previously hypothetical concerns that GM crops could unintentionally spread and take over traditional forms of agriculture. Moreover, maize is the species that companies use for much of their research into further uses of biotechnology, including 'growing' pharmaceutical compounds using crops. As such, the concern that GM strains could accidentally spread to non-GM crops and contaminate them with a pharmaceutical compound implies potentially serious health and safety problems.

The controversy also gathered momentum because Chapela had been leading a fight against a controversial research partnership between the biotechnology firm Novartis (now Syngenta) and the University of California at Berkeley, which gave the company privileged access to the university's plant scientists. Novartis agreed to provide up to $25 million over five years in return for a role in handing out the money and rights to the research findings. Chapela's struggles became a symbol of the erosion of academic independence from corporate influence. This even extended to his tenure case.[19] Chapela came up for review in September 2001, and received overwhelming support from his colleagues. The college's acting dean approved their decision, and then a campus-wide tenure-review committee voted unanimously in Chapela's favour; a mere eighteen months later the campus budget committee and the chancellor of the university voted to deny Chapela tenure. After international protest and several grievances and lawsuits, in May 2005 a new chancellor of the university finally approved Chapela's tenure.

Again, the scientific conclusions were only part of the story. The ecological and agricultural consequences of the contamination that Chapela and Quist reported are worrisome for some. The concern that GM crops could surreptitiously find their way into conventional crops raises concerns about environmental contamination, genetic drift and agricultural sovereignty. For many, industry has yet to establish the environmental impact of GM crops. Another concern, that industry has yet to demonstrate the safety of GM food for human consumption, played out during another scandal: the StarLink controversy.

StarLink is a GM variety of corn that produces the Cry9C protein. Government agencies only approved the corn for use in animal feed, not for the human food supply, because of concerns it might trigger allergic reactions. In 2000 food manufacturers accidentally introduced StarLink corn into several food products that found their way onto supermarket shelves. This triggered a recall of 300 brands of taco shells, chips and other American foods. The controversy forced Kellogg and ConAgra to shut down production lines for almost two weeks to make sure there was no StarLink in their systems. Tyson Foods Inc., the world's largest poultry producer, even refused to buy StarLink for their feed as the controversy grew. Beyond the immediate financial issues, some believed the StarLink case was simply a harbinger of more troubles to come. Not only were the varieties of corn not separated, it took the third-party consumer group Friends of the Earth to test the products to discover they had been tainted. This was an example of how government regulations and industry procedures had failed to keep a GM product from the food supply.

In November 2012 Dr Gilles-Eric Séralini and colleagues at the University of Caen in France published a heavily criticized study of the effects of GM maize and the Roundup herbicide in *Food and Chemical Toxicology*. As part of the pre-publication embargo, the authors took the unusual step of forcing reporters to sign a non-disclosure agreement that barred them from seeking outside sources. Upon publication, groups that opposed GM crops or industrial agriculture promoted the study's findings: that rats fed either genetically engineered corn or the herbicide Roundup had an increased risk of developing tumours, suffering organ damage and dying prematurely. Shortly after publication, several scientists questioned the findings, citing statistical and methodological problems. Commentators felt that Séralini's embargo was a deliberate manipulation of journalists to ensure that the first news stories published about the study would not be critical of its methods or results. In December 2013, after a lengthy series of letters to the editor, both pro and con, the journal officially retracted the publication, despite 'no evidence of fraud or intentional misrepresentation of the data'. This retraction set a precedent for a peer-reviewed publication that has serious implications for scientific inquiry.

Scholarly and scientific literature is filled with controversial papers that evoke criticism and arouse passionate debate. Neutral scientific inquiry is done by people who want to investigate something, not by those who want to advocate for a particular point of view. The publication, retraction and resultant media attention brought issues of corporate influence and impartial advocacy to the fore. The belief that scientific inquiry is a neutral, impartial arbiter, free of conflicts of interest, is integral to public trust in science. But this trust is harmed by efforts, real or imagined, to suppress scientific findings. The attempt to create unreasonable doubt around inconvenient or unwelcome results jeopardizes public confidence in scientific methods and institutions. No matter the merits of a particular study, unless everyone trusts the process, neither the public nor the scientific debate will abate. Independent scientists devoted to the public interest and professional integrity are fundamental to an honest, rational and truly scientific debate surrounding new technologies like genetic modification.

These scientific controversies, involving researchers working in the United States, England, Mexico and France, make it evident that a multitude of stakeholders are influencing discussions about genetic modification. Some people criticized these researchers for using questionable science to advance personal agendas. Food manufacturers, government agencies, environmental groups and other social actors hijacked scientific arguments about GM food to serve other motives. In this case, as with other controversies, contested interests and symbolic battles characterize scientific judgements and evaluations over claims of expertise.[20] Multinational biotechnology corporations use their power to challenge the scientific authority of those who question their products.[21] In turn, opponents of these corporations, of the industrialization of agriculture, of U.S. policy and of globalization have found a common rallying point with environmental and consumer advocacy groups.

Given past experience, it would be wrong to assert that the controversies posed by GM food can easily be resolved through scientific education or better public understanding. The science of genetic modification, as with many modern technologies, is sufficiently advanced to be considered akin to magic by many of us. The general public may not possess 'expert' knowledge, as traditionally defined, or even a great deal of scientific understanding, but this does not mean that they have nothing to contribute to decisions about science and technology. In particular, GM food brings a moral debate to the forefront. Not only do scientists tamper with nature in a way that many believe was previously reserved for a divine force, they do so with the assistance and protection of government regulation. The moral tinge of this fundamental question of what is 'natural' brings forward profound questions that science is ill-equipped to handle. To understand these issues requires an examination of the food ecology

– the key stakeholders, including industry and its related organizations, academia, government, advocacy groups and various interested publics.

Differences in public perceptions, interest group dynamics, political systems and industrial structure have driven European and u.s. agricultural biotechnology policy in opposing directions.[22] In the United States, technology firms and large farmers have pushed for and obtained comparatively permissive regulatory standards; in the European Union, advocacy groups have urged highly precautionary regulation of GM food.[23] The controversies surrounding genetic modification have become largely symbolic in content. Rather than trying to address the challenges and opportunities of GM food, stakeholders appropriate public trust. In this context, public trust is a resource, like money and political power.

5

GETTING BACK ON TRACK: THE TENSION BETWEEN IDEALISM AND DOOM

David Foster Wallace wrote, 'For practical purposes, everyone knows what a lobster is. As usual, though, there's much more to know than most of us care about – it's all a matter of what your interests are.'[1] We could say the same thing about our food supply in general, or about GM food specifically. The ubiquity of food-related television programming, as well as food- and restaurant-themed best-selling novels, has seemingly stoked interest in all things food-related. This interest will cause some people, like the readers of this book, to further explore some of the big challenges facing our food system. Hopefully, some will propose intriguing solutions – or stimulating questions, for that matter – that may force us to rethink the status quo. However, I think that the most important thing we can do is to resist the temptation to focus on the negatives about GM food. Instead, I propose that, while keeping alert to the controversies and potential problems, people should use their available time to promote new options for addressing critical issues affecting our food system.

The demands of a globalizing economy and the increasing expectations of consumers worldwide will continue to place new and ever-growing pressures on producers, consumers and governments. In particular, consumers want the current controversies resolved. Despite some loud voices, however, the controversies surrounding GM food are closer to momentary turbulence than an existential crisis that would force stakeholders to reconsider or end their commitment to genetically modified crops in the global food system. Moreover, science and technology, even when applied to food, are moving targets. For example, mutagenesis – treating seeds with chemicals or radiation to cause random mutations – has been used for decades without much public notice; Canada is the only country that regulates the resulting crops. But recent reports are drawing attention to the practice, and even scholarly groups like the U.S. National Academy of Sciences are raising concerns.[2] Similarly, the controversies and worries about GM food may give way to unknowns related to nanotechnology or synthetic biology, both of which may make genetic modification seem tame by comparison.[3] Or, perhaps controversies about GM food will wax and wane, much like concerns over nuclear power.

To put the narrative of GM food controversies in perspective, let's take a look at a brief summary of nuclear power's historical trajectory. For much of its history, the anti-nuclear-power movement has had limited success in forestalling the spread of nuclear power worldwide, much as the anti-GMO movement has failed so far to end or slow the spread of genetic modification within the global food system. Starting in the late 1970s, protesters staged many large anti-nuclear demonstrations and protests, particularly in Europe and the United States. These events brought attention to the proliferation of nuclear power plants, but did not slow their adoption. Many governments did not pause their nuclear development plans and rethink their energy policies until disastrous accidents occurred, such as the Three Mile Island accident in the United States in 1979 and the Chernobyl disaster in the Ukraine in 1986. There are no equivalent catastrophes involving genetic modification. But even if there were, our reliance on genetic modification still might not end. Soon after the previously mentioned 1979 and 1986 disasters, the nuclear industry advocated for nuclear power as environmentally friendly, efficient and safe due to advances in the design of nuclear reactors used for alternative sources of power.[4] As this narrative started to unfold and popular opinion began to be swayed, the Fukushima Daiichi nuclear disaster occurred in Japan and revived anti-nuclear sentiments worldwide in 2011. Globally, more nuclear power reactors have closed than opened in recent years.[5] As a result, *The Economist* issued a special report that called nuclear energy 'the dream that failed'.[6] Given the limited influence of genetic modification in addressing global food issues, the GM industry could be a dream that fails, too. Still, knowing the vicissitudes of global science and technology policy, as seen in the continued reliance on genetic modification technology, the nuclear dream may be revived again, perhaps in China or India.[7]

The example of nuclear power helps illustrate how the ability to limit potential catastrophe is greatest at the point when societies are choosing whether to start using the technology.[8] After the technology is in use, even major catastrophe does not mean use of the technology automatically stops. In applying this framework to food, experts, scientists, industries and governments usually take the lead in constructing the menu of choices available to us.[9] Once they have broadly adopted a particular technical solution, longevity and inertia ensure that even a major calamity cannot wipe it from the world. Once these groups decided to move forwards with GM food, the non-expert public's choices became limited to influencing how it could be regulated and, to a limited extent, where and when it could be produced; the public could no longer influence whether or not GM crops were available.

In general, agricultural techniques mature and replace each other over decades. Even though farmers and agribusiness have rapidly adopted GM seeds, that does not mean it is a permanent change. The public's limited choice will persist if these influential groups continue to focus on GM food as the signature food issue. However, when our primary concern is whether we use genetic modification or not, we neglect any alternatives to items on our current menu of food choices or food system practices. GM seeds do not have to, and likely will not always, dominate the industry.

The modern food system of the last fifty years has emphasized a mechanized, centralized and globalized infrastructure that has managed to produce enormous amounts and varieties of food available at our convenience. This type of infrastructure and mass production of different foods means that the production of GM food requires the functioning of a complex, interdependent web of stakeholders who must, in turn, cooperate with, compete against and trust one another. This highly technical and scientifically complex organization of the global food supply is characteristic of a 'risk society', in which the public requires a promising pool of trust to function adequately.[10] In this scenario the public must trust countless companies, experts and regulators, even though they each have their own vested interests. Given the intricate supply chains for all of our food, the public has to trust that those who have a specific task or responsibility will perform their duty in a way that people can count on. As a result, the public relies on experts, even though experts will sometimes fail. Moreover, it is increasingly difficult to know who is really the right expert for the job.[11] If nothing else, these human uncertainties put social and political issues at the core of debates surrounding the risk of technical solutions to our modern agricultural problems.

As the role of the farmer has been reduced to producing commodities for large agribusiness corporations, various aspects of food and its production have become more obscure. For instance, though people throughout the world purchase and consume GM food every day, they do not understand the science behind it. Moreover, given that manufacturers do not always label genetically modified foods as such, people are not forced to make reasoned, well-informed or explicitly thought-out decisions to purchase or avoid them. While a person's role as a consumer in a grocery store highlights individual, economic interactions, the social and cultural expectations about safety, quality, taste and nutrition, as well as ethical aspects of food production and distribution, are also embedded in these interactions. Therefore, it is important to consider the supra-individual aspects of buying food in order to understand all elements that could be addressed for improving our food system, as opposed to simply supporting or rejecting genetic modification.

With a highly differentiated division of labour, where social exchange takes place across long physical and social distances, people delegate the responsibility for meeting our needs to others, often to organizations represented by a chain of strangers and agents with expertise. These organizations and experts produce, process and distribute food; they also provide knowledge and information about our food. The public relies on these actors because most people simply cannot collect, process and interpret all relevant data themselves. The public has to rely on the representations and assessments of experts.

We have several reasons to trust our food supply, including past experience with specific brands and shops, quality indicators, traceability and regulatory assurances of all kinds. Still, partly due to the generally anonymous character of the food exchange, there is an opportunity for consumers, stakeholders and experts to put their self-interest above all, despite the risk of negative consequences for others. In these situations, guardians and watchdogs for the public interest may relieve some of this uncertainty with whistleblowing, regulatory procedures or relevant research. While government agencies often fulfil this function, other entities assume this role on occasion; the media, public advocacy and consumer protection organizations on the one hand, and experts and scientific institutions on the other. Even though these actors may share a goal of safeguarding consumer interests, it may not always be their priority. Capitalistic incentives and the potential for individual actors' goals to conflict, all within this complex system of checks and balances, make it difficult to know whom to trust. Moreover, despite experts on the job, the indeterminate nature of scientific knowledge and inquiry produces uncertainty, and unintended consequences for people and the food they consume.

For example, one of the promises of using GM seeds is that farmers can use specific herbicides, like glyphosate, and conservation tillage methods for weed control rather than an array of herbicides. Conventional tillage is when a farmer completely turns over the soil to kill weeds and incorporates plant matter or manure as fertilizer. Conservation tillage – whether it is no-till, ridge-till, mulch-till or chisel-plough – disturbs the soil less and leaves at least 30 per cent of the previous crop's residue on the soil surface at the time of planting. Anything that farmers do to decrease erosion losses also decreases the need for chemical fertilizers and amendments because the topsoil generally contains the most organic matter. Moreover, cropland erosion chokes up streams with phosphorus, which is a major cause of freshwater pollution. In this sense, when farmers adopt these crops along with conservation tillage, they indirectly benefit the environment by reducing soil losses. So conservation tillage can have some real benefits for industrial agriculture.

However, it turns out that the adoption of conservation tillage coincides with the incidence of harmful algal blooms. Without conventional ploughing, the phosphorus in the manure and chemical fertilizers that farmers apply to their fields becomes concentrated in the surface layers of the soil. So even though, with conservation tillage, the run-off and soil erosion is reduced, the high phosphorus concentration at the soil's surface stimulates algae growth, creating an imbalance that affects fish populations and increases toxin-producing microorganisms. Moreover, these algal blooms are exacerbated by heavy rainfalls, which are predicted to become more frequent as climate change proceeds. So in a sense, farmers have not so much solved a problem as they have exchanged one set of problems for another.

Market competition also complicates the priority of consumer interests and limits potential for an equalized checks-and-balances system. Stakeholders compete against one another, making claims to support their position. When the public supports their claims – whether rhetorical, scientific, political, social and so forth – this represents a potential increase in that stakeholder's power. Power has to do with the ability of one group to reach its goals, despite another group's opposing interests.[12] As the political scientist Robert Dahl famously noted, it is difficult to posit the existence of highly concentrated forms of power without sounding paranoid.[13] However, understanding the different types of power that companies can exercise is instructive for critically assessing companies' impacts on our food system.

In the first type of power, there is equal access to decision-makers, but one group has more resources or skills than the other. For example, supporters of a California ballot initiative to label GM food raised almost $9 million, but were outspent by their opponents who spent $46 million, including $8 million from Monsanto alone.[14] So, the group with more money beats the group with less. In the second type, power generates decisions and causes 'non-decisions'. One group beats another by preventing particular political decisions from becoming options for consideration, one of many strategies for controlling issues on the political agenda. If agribusiness has so much control that the anti-GM activists may view slight changes as impressive victories, including even a decidedly ambiguous trade agreement on the process by which countries can refuse to accept GM products, then the significant non-decisions – like whether GM food should be in the food supply at all – are ignored or accepted as given impossibilities. Finally, in the third aspect, one group cannot even conceptualize something beyond the false consensus that serves the other group's interests. In other words, by now, the opponent views the previous defeats as unremarkable and normal. Most critics of genetic modification do not propose an end to large-scale agriculture; rather they concentrate on whether genetic modification is an agricultural technology that will benefit the public. A small part is critiqued instead of the entire machine, so to speak. In each of these dimensions it would seem that agribusiness has all the power in our modern food system.

Competing paradigms, such as sustainable agriculture and organic agriculture, do not directly match or threaten the power of modern agribusiness and do not imply the end of large-scale farming. If these alternatives tried to produce food on the same scale as the current agricultural paradigm, they would need to address the same problems of scale, geographical distribution, regulation and market control as the conventional food system. This serves to reinforce the power of the incumbent systems of large-scale corporate agribusiness that have already addressed these issues, albeit imperfectly. For example, when Wal-Mart

Stores Inc. – the largest American grocery retailer – decided to embrace organic products, food manufacturers like the Kellogg Company, PepsiCo Incorporated and General Mills made plans to follow this decision and introduce organic products into their lines, necessitating large-scale production and distribution.[15] Even when organic agriculture is privileged in the food production world, it fails to challenge the modern system, serving instead as one of its profit-making strategies.

The general problem still remains: the relations between power and the interests and responsibility of actors are not easily detected or even visible. Because merchants and food manufacturers make the initial decisions, set the initial agenda and define the 'appropriate' frames for emerging food-related technologies like genetic modification, they are structurally and financially powerful when it comes to agricultural issues. In this position, they can manipulate debates to their benefit, even if at the cost of others. For example, though it seems paradoxical, food manufacturers might benefit from debates about the best way to label GM food because the debate assumes that everyone has already agreed to sell GM food. This suggests that anti-GM consumers and advocacy groups should instead spend their social capital debating whether farmers should grow GM crops at all. Alternatively, if debates were to centre on the cost-efficiency and production yields of GM crops versus conventional farming practices, then discussion about the ethics of patenting staple food crops would be lost or relegated to conversational asides. As power players and their decisions remain invisible, so do other food topics of importance.

Excessive advocating for or against a particular idea, as we see from the current pro and con debates over genetic modification, makes it easy to forget that other options exist for improving our food system. This phenomenon sometimes serves a positive function. For instance, parents quickly learn to offer their toddlers a restricted choice – 'do you want to wear the blue shirt or the red shirt today?' – to elicit cooperation. By offering pre-approved choices, we guide our children to an acceptable decision and save ourselves from the seemingly limitless search for alternatives. But more than that, by restricting choices, other options are not made visible. This happens, for example, when the media report polls that measure approval and disapproval of genetic modification. These polls are easy to interpret and serve as tempting fodder for media pundits and opportunists with a vested interest. Saying that a large percentage of the public wants or uses some particular technology makes for a wonderful sound bite. But assessment of a technology is not that straightforward. It is important to note what information is communicated and what information is left out. These opinion polls are superficial, almost to the point of meaninglessness, because they focus on a narrow piece of information without providing proper context or background information about the entire issue at hand. Additionally, the media source may not include details of how the poll was conducted, which prevents the public from knowing any of the limitations to the poll's validity. These polls are often best left as exercises in survey methodology and examples of how question framing and word choice can sway public response.[16] But even if we do away with publicizing the poll results, their sound bites alone can have dangerous effects.

Suppose that polling on genetic modification indicated a clear majority in support; the sponsoring company and the associated regulators might assume there is little work to be done. For marketers, the support might be taken as a measure of success, either commercially or in shaping public sentiment. Those who oppose genetic modification might use these numbers as a rallying cry for increasing awareness, hoping to weaken support in the next round of polling. But a more nuanced view would help the public deliberate the complexity inherent in their choices. Deepening engagement rather than manufacturing quick consensus provides less provocation for partisans and more reassurance for the public. In addition to the publicized health and environmental debates, a more subtle but important part of what makes GM food so controversial is this lack of meaningful engagement and the continued provocation of partisans on both sides.

These controversies surrounding GM food are about social and political power, democratic practice and corporate responsibility as much as they are about the global food system. Even if genetic modification gains worldwide acceptance, it is a marker for industrial, processed, United States-dominated agriculture and McDonaldized values. This means that many opponents are using the issues related to genetic modification as a surrogate for a values war against American powerhouse companies they do not like and do not trust. They focus their efforts on GM seeds because they are the clearest and most obvious manifestation of an aggressive privatization of what many consider a fundamental human right to food and nourishment.

So if agribusiness loses the struggle, the losses would not only be economic, but politically devastating. The losses would reflect agribusiness's loss of both power and public trust. However, I do not believe agribusinesses will lose their power. All they have to do to retain power and trust is to be minimally competent, act with some concern for others and help the public navigate some of the uncertainty inherent in new food science and technology. Because their positions are structurally secure and they already hold so much power, it would be incredibly difficult for them to lose their dominant position. It would also be incredibly difficult for another group to supplant them. Multinational agribusiness companies are obviously key stakeholders in how science proceeds, occupying a crucial position between acceptance and uncertainty. More directly, they assume social and ethical responsibility by creating and circulating knowledge about GM food. Some of the controversy would subside, as well as any potential threat to their power, if they clearly communicated that genetic modification is not the panacea that its proponents claim.

However, genetic modification is not the biggest challenge in our food system either. Climate change will bring new challenges for agriculture, as more extreme weather conditions are expected to negatively impact yields.[17] The future is likely to include rising temperatures that lengthen the growing season and potentially increase production in some areas. Extreme weather events represent an unheralded level of risk and volatility for farmers. For example, China traditionally produces more corn than any country in the world, except for the United States. However, excessive rain damaged the corn crops in northern China in 2014; in 2011 drought conditions in China turned the country into a net importer of grain.[18] While farmers have always had to contend with some variability and volatility from year to year, the violent ups and downs of catastrophic weather events that have vexed agricultural producers in the last few years are a harbinger of the future. Even the intensified crop yields that have happened to date are not sufficient to meet the projected demand for food by 2050, and in some places crop yields have already stopped increasing.[19] Eroded, degraded nutrient-poor soil, inadequate and polluted water supplies, soaring energy costs, international trade agreements and policies that alter the supply and demand for commodities, and excessive food waste at each step in the supply chain are just a few of the major issues that we all face, and that are beyond the reach of a single technology.

Reframing the controversies around GM food as a challenge to revolutionize our agriculture system can open up new avenues for discussion. This new focus encourages a more constructive conversation about the role that science and technology can and should play in the future of the global food system. Challenges can be met if we are technologically agnostic, denying and doubting the possibility that we hold the ultimate knowledge of what is best. In other words, be humble and be open to the possibilities. Our agricultural system is not truly wed to a future built on genetic modification. Right now, that is simply the technology that allows for the easiest path and the highest profits. However, agribusiness firms will happily switch to an easier or more profitable path if presented with one. In fact, they may have already started moving on to less controversial though no less problematic methods, such as mutation breeding and marker-assisted selection.[20] If business opportunities exist outside of a genetically modified future, the modern food industry will gladly take on that challenge.[21] The efficiency and the power of science and technology are important tools when thinking about the challenges in our food system.

Something that we will need to consider is how we can let scientists use their tools and talents to help us achieve shared goals. We should replace our technocratic utopian hubris with more humble expectations and understandings about what scientists can control and predict in advance. Even if scientists and industry think GM food is not a problem, the process itself does not have to remain a priority; we do not have to develop everything just because we can. But this also means that critics should not fundamentally reject potential technological solutions to current challenges. If activists hope to blunt the worst impacts of our modern food system, Monsanto and DuPont might make for unlikely but remarkable allies because of their monopoly of some potentially groundbreaking technology. Starting with chemicals, then biology, and now data, they continue to show the public that another potential industry-wide transformation is perennially at hand.

Agribusiness companies, like Monsanto, DuPont, John Deere and other power players, have started to invest billions of dollars into aggregate precision farm data that combines superaccurate GPS receivers with harvest, planting and soil data to formulate specialized seed-planting recommenda-tions.[22] They want to accelerate, streamline and combine all the data from farm equipment, such as combines and tractors, with highly detailed records on historic weather patterns, topography and crop performance. This technology could help optimize plantings, as long as farmers trust the companies not to misuse the data, and the companies honour that trust. It is hampered by the fact that a surpris-ingly high percentage of farmers across rural communities, even in developed nations like the United States, are not con-nected to a broadband and communication infrastructure – or if they are, their network is unreliable or limited.[23] The lack of a dependable and robust communication channel for all of this data means this does not amount to a widespread solution. But by forgoing a universal remedy to all of our agricultural ills, we can become more nuanced and plural-istic in our approaches, embracing all forms of agricultural inventiveness.

By focusing on longer-term, public investments in agricul-tural research we may avoid some of the more pernicious aspects of the agricultural treadmill. As Michael Dimock, president of Roots of Change, wrote: 'farmers must rapidly sequester more carbon, eliminate broadly toxic and long last-ing pesticides, pasture more animals, build soil health and increase crop and livestock diversity.'[24] Here, too, we should keep in mind the influential psychologist Abraham Maslow's golden hammer metaphor.[25] Technology may not always be the best answer, but sometimes there may be nail-like prob-lems that would be matched with hammer-like technological solutions. The parts are constantly changing; so, too, will the necessary and acceptable conditions for a global food system need to be ever-changing to help reach minimally controversial solutions to bigger problems – engaging in finding common goals and a common good is a start. Anecdotal evidence strongly suggests that while some actors believe public engagement is an inherent good, many within the scientific and policy communities see it primarily as a way of informing the public and deflecting criticism, rather than as a way of incorporating public values and preferences into the policy process. While these alternative conceptualizations can and do coexist, they represent conflicting views of the role of the public. Based on changing agribusiness oratory, it would seem that industry has realized, even if belatedly, the value and importance of consumer preferences and desires. It would be easier to be enthusiastic about agribusiness rhetoric, however, if it acknowledged that social actors in the food system should take on social responsibility alongside their appetite for profit. One of the key responsibilities should include recognition that although the food system is global, agricultural solutions should emphasize local cultural sensitivity.

For those who oppose industrial agriculture as a social ill, organic agriculture may seem a respectable cure, but it is more of a niche in the dominating food paradigm in terms of scope. That is not to say that organic agriculture is a failure, but it also may not be suited as the main priority for our present food system. Similarly, agritourism that capitalizes on the current trendiness of farms, farming practices and the pastoral aesthetic belies the realities of current large-scale farming and production. This romantic stereotype should be fought against and abolished if it in fact is forcing us to ignore current farming issues, taking away money and attention from a more sustainable model of local farming. Critics are right to point out that this is a quaint and dated image. These images and practices are not, and cannot be, the only solution for a broader goal of sustainability in light of large corporate farming issues and the realities of local farming in general. Farming is not just a communion with soil and water alone.

It sounds like a platitude to say that I think the future of our food system should balance our need for food with our environmental impact while taking care of our cultural, social and economic needs. But this would not be an easy task, even if we just wanted to satisfy the needs of a small nation. Creating a global food system that accounts for all these factors is daunting and may lead to solutions that are less satisfying than 'satisficing'. The Nobel Prize-winning social scientist Herbert Simon coined the word 'satisficing' to explain the tendency to select the option that seems to address the most needs rather than the optimal solution.[26] This way of making decisions can lead to unsettling trade-offs. For example, people who have prioritized locavorism, animal welfare, reduced pesticide use or any number of specific agricultural practices often disregard the human labour abuses in the system. When our primary concern is what we put into our bodies, we neglect the bodies that make our food available. Though this may sound pessimistic or judgemental, it should be empowering and enlightening. Though entrenched, the current food system can be changed. Awareness of all areas in need of improvement in our food system, and the barriers to their improvement, is necessary for people to address them. People construct systems and their problems, and therefore have the power to change them.

That being said, complex systems pose complex challenges that are not easily solved. If we are trying to balance several, sometimes competing, needs, pressures from consumers are likely to sway reforms. By shifting trade-offs in this sort of system, the driving questions and controversies then become how incrementally or how radically the consuming public wants to reorganize the entire food system. Sometimes, drastic reorganization of a problematic system might not seem worth it if it means giving up a valued advantage from the system. For example, though some romanticize small-holder farms, large-scale farming actually helps blunt the physical and economic hardships, as well as the poverty and undernourishment, pervasive in small-scale agriculture. As the historian Rachel Laudan has declaimed: 'If we romanticize the past, we may miss the fact that it is the modern, global, industrial economy . . . that allows us to savor traditional, peasant, fresh, and natural foods.'[27] However, I cannot see myself promoting the overly narrow focus of the current mainstream agricultural system. Its focus on yields, regardless of context or culture, is drastically out of balance. It routinely determines and extends pockets of advantage, and need, around the world. Given its global market focus, it is unsuited to adapt to local cultures and processes.

The honest answers are not sexy and are devoid of conspiracy theory intrigue, relying less on advanced science and technology and more on applying what we already know. The current obsession with genetic modification crowds out more important issues. In fact, the lasting impact of genetically modified technology may be the industry-wide consolidation that was made possible by its profits. Fewer and fewer firms control the seeds, herbicides and pesticides as well as the precise agricultural data that are part and parcel of modern farming. The sad truth is that the biggest problem with our agricultural system is a social and cultural reluctance to refine our current approaches: we prefer to endlessly and perhaps fruitlessly search for a new, all-encompassing approach. Agribusiness companies tend to love solutions that they can profit from, and those solutions currently take the form of scientific hegemony enforced by intellectual property law. Moreover, there exists a technocratic impulse to tailor problems to fit solutions, instead of the reverse. Additionally, the United States' incessant belief in superior technology takes on a colonial tinge. The relentless promotion of this belief makes many feel like proponents are claiming that u.s. approaches to agriculture are superior and universally applicable, despite social, cultural, economic and political differences across societies worldwide.

The social and cultural tendency to apply a one-size-fits-all technology that I have described applies to the specific example of GM foods. Even on its own terms, the promise of GM foods to create abundance is an empty promise. Even when yields have increased, the sponsoring firms' narrow technical focus has yet to pave ways for distributing the resulting crops more equitably. Increasing the production of food crops does not address problems of inadequate purchasing power and can, as the Nobel Prize-winning economist and philosopher Amartya Sen's analysis of famine shows, exacerbate inequalities of wealth and power.[28] People are not hungry because there is no food, they are hungry because there is no entitlement to food.

The United Nations projects that a 60 per cent increase in agricultural production is needed to help feed a projected population of nine billion by 2050.[29] This figure assumes we will go on wasting food at the current rate and continue to consume richer diets as incomes increase. But neither of these things needs to hold true.[30] Reducing waste can increase supply without any of the ecological downsides of growing more. The UN estimates that 1.3 billion tons of food produced globally, representing as much as one-third of all food, is wasted each year for preventable reasons, including insufficient markets, poor storage and disposal by processors, distributors, restaurants and consumers.[31]

In order to resolve the issue of excessive waste, we might be inclined to look to technology for the answers. But in regard to power structures, introducing powerful new agricultural technology into deeply unjust rural social systems simply reinforces the injustices within those systems.[32] Conceivably, introducing powerful new agricultural technology into a relatively equitable and secure system *might* help the poor, but there are easier answers than reliance on advanced technology.[33] Large improvements in crop yields are possible in many developing countries through a sustained focus on adding organic matter, small doses of fertilizer and water-efficient irrigation systems.[34] In other words, minimize the use of technology and take less risky approaches such as attending to soil and water management. This should be a fundamental step, regardless of which direction we choose for our future agricultural systems. Fortunately, many sustainable agriculture practices such as this should also result in higher resilience, meaning we can help protect against yield losses while simultaneously improving environmental performance.[35] This does not solve every problem. But starting in the right place improves our chances of ending up at the right destination.

I have not intended to write a polemic simply in favour of a particular agenda for social change. Instead, I have tried to look at some of the controversies surrounding GM food and find lessons we can use going forward. Distilling these lessons into something worthwhile means distinguishing how the world is and how we might like it to be. After more than a decade of research, I can state that I hold a relatively neutral position on GM food. Though the agribusiness industry's lack of support for more research on potential human health threats troubles me, I do not have immediate, grave concerns based on my understanding of the scientific, medical and scholarly literature. This is not to say there is no potential threat; rather, I do not believe that GM foods pose a more obvious health threat than other practices. Their potential environmental threat concerns me more, but the causal links seem more directly tied to the ways that large-scale farming practices react to incentives for giant monocultures. This same system of vast monocultures hastens herbicide resistance and often neglects more holistic practices that would benefit farm health in the long term. Moreover, I am concerned that GM crops have continued to push more of the world towards industrialized production, with its reliance on higher water, fertilizer and pesticide use. In this sense, genetic modification as applied to agriculture has been an answer in search of a riddle, ignoring and exacerbating larger agriculture issues in favour of profitable problems.

However, it should be clear by now that I do not want my neutral position to be mistaken for support. While future GM crops could add other beneficial plant traits, which might help boost productivity in crucial crops, I think the best answers lie elsewhere. As the Pulitzer Prize-winning novelist Bernard Malamud wrote, 'if your train's on the wrong track, every station you come to is the wrong station.'[36] Technocratic solutions do keep leading to the wrong station. Sure, they might stop at an interchange or allow us to transfer gracefully at some point. Science and technology might even help us find a faster route to ending some of our agricultural woes. But right now, our global food problems do not stem from lack of agricultural technology or of scientific know-how. In large part, they result from unequal land distribution based on domestic and international economic and trade policies that favour powerful, multinational corporations in their defence of expansive intellectual property rights. Additionally, global food problems also stem from the ecological, cultural and social devastation wrought by solutions that do not give enough deference to context.

Agribusiness firms increasingly use science, technology and unprecedented access to genomic resources to change agriculture. Each of their advances will bring more controversy that science alone is ill-equipped to handle. If we continue to focus narrowly on scientific advances enveloped in controversy, the struggle to introduce, discuss or implement alternatives will only intensify. The real controversies posed by genetic modification cannot be easily or quickly resolved through scientific education or better public understanding. The application of new technologies brings a variety of social, political, cultural and moral debates and controversies to the forefront, requiring time and energy that could be used instead to address problems that can be resolved without relying solely on cutting-edge science.

This book has allowed me to write about the environment, decision-making processes, culture, acceptability, organizations, risk and social action in the context of food biotechnology. For more than ten years, I have been fascinated with the controversy surrounding GM food as a proxy debate for broader issues of social and political power, cultural values and corporate responsibility. I have found that there is no universal recipe or set of levers for agribusinesses, regulators, activists and experts to agree on for creating a more perfect global food system. A universal formula for a global food system would rest on the flawed presumption that stakeholders themselves do not change. Competition between stakeholders, and individuals' ever-changing relationships – with science, agriculture and regulators – preclude a fixed, permanent formula. A noted scholar, Juliet Schor, has remarked that the twentieth century was characterized by the search for one right way, but that the present and future are best characterized as an ecosystem of solutions and many right ways. That there is no singular recipe suggests a broader conclusion.

In the end, the controversies surrounding GM food reflect the irreducible social vulnerability and undeniable uncertainty brought about by our modern, global system of agriculture. My bottom-line recommendation is not a rejection of GM food as dangerous or inherently flawed. Instead, I contend that what is inherently flawed is the industrial system of agriculture's prescribed path that promotes GM food. In the end, the future of our food supply may be shaped by cultural and social innovation and cooperation as much as, if not more than, by a series of discrete scientific discoveries. There will always be controversy. But, globally, we will be better off using our knowledge, money, research, media exposure and public advocacy to find better solutions that balance environmental, cultural, social and economic needs.

Although some of the tension in global agriculture may not be necessary and much of it may not be polite, it can be creative. As Hunter S. Thompson wrote, 'It was the tension between these two poles – between a restless idealism on one hand and a sense of impending doom on the other – that kept me going.'[37] We, too, can use this continuing tension between restless idealism and an impending sense of doom to keep us all searching for better solutions for our global agricultural system.

REFERENCES

INTRODUCTION: GENETICALLY MODIFIED FOOD: REMAKING THE GLOBAL FOOD SYSTEM

1 William J. Broad, 'Useful Mutants, Bred with Radiation', www.nytimes.com, 28 August 2007; 'Texas Grapefruit History', www.texasweet.com, accessed 8 July 2014.

2 Stanton B. Gelvin, 'Agrobacterium in the Genomics Age', *Plant Physiology*, CL/4 (2009), pp. 1665–76. Shyamkumar Barampuram and Zhanyuan J. Zhang, 'Recent Advances in Plant Transformation', in *Plant Chromosome Engineering*, ed. J. A. Birchler (New York, 2011), pp. 1–35.

3 Agnès E. Ricroch, Jean B. Bergé and Marcel Kuntz, 'Evaluation of Genetically Engineered Crops Using Transcriptomic, Proteomic, and Metabolomic Profiling Techniques', *Plant Physiology*, CLV/4 (2011), pp. 1752–61.

4 'ISAAA Brief 49-2014: Executive Summary', www.isaaa.org, accessed 4 February 2015. See the World Bank for region size data, 'Land Area', http://data.worldbank.org, accessed 29 July 2014.

5 'ISAAA Brief 49-2014: Executive Summary'.

6 Ibid.

7 Marion Nestle, *What to Eat* (New York, 2006).

8 William K. Hallman et al., *Public Perceptions of Genetically Modified Foods: A National Study of American Knowledge and Opinion* (New Brunswick, NJ, 2003).

9 Robert Falkner, 'The Political Economy of "Normative Power" Europe: EU Environmental Leadership in International Biotechnology Regulation', *Journal of European Public Policy*, XIV/4 (2007), pp. 507–26.

10 George Gaskell et al., 'Troubled Waters: The Atlantic Divide on Biotechnology Policy', in *Biotechnology, 1996–2000: The Years of*

Controversy, ed. G. Gaskell and M. Bauer (London, 2002), pp. 96–115.

11 Paul Lewis, 'Mutant Foods Create Risks We Can't Yet Guess', Since Mary Shelley', www.nytimes.com, 16 June 1992.

12 Rachel Schurman, 'Fighting "Frankenfoods": Industry Opportunity Structures and the Efficacy of the Anti-biotech Movement in Western Europe', *Social Problems*, LI/2 (2004), pp. 243–68.

13 Timothy C. Earle and George T. Cvetkovich, *Social Trust: Towards a Cosmopolitan Society* (Westport, CT, 1995).

14 Wendell Berry, 'The Pleasures of Eating', in *Cooking, Eating, Thinking: Transformative Philosophies of Food*, ed. D. W. Curtin and L. M. Heldke (Bloomington, IN, 1992), pp. 374–9.

15 United States Government Accountability Office, *Genetically Engineered Crops: Agencies are Proposing Changes to Improve Oversight, but Could Take Additional Steps to Enhance Coordination and Monitoring* (Washington, DC, 2008).

16 Steven Mufson and William Branigin, 'European Union Urges Testing of U.S. Wheat Imports for Unapproved Strain', www.washingtonpost.com, 31 May 2013.

17 Randy Gordon, 'NGFA Estimates Up to $2.9 Billion Loss to U.S. Corn, Soy in Aftermath of Trade Disruption with China Over Detection of Unapproved Syngenta Agrisure Viptera™ MIR 162 Corn', www.ngfa.org, 16 April 2014.

18 Scott D. Sagan, *The Limits of Safety* (Princeton, NJ, 1993).

19 Charles Perrow, 'Organizing to Reduce the Vulnerabilities of Complexity', *Journal of Contingencies and Crisis Management*, VII/3 (1999), pp. 150–55.

20 Kristen Purcell, Lee Clarke and Linda Renzulli, 'Menus of Choice: The Social Embeddedness of Decisions', in *Risk in the Modern Age: Social Theory, Science and Environmental Decision-making*, ed. M. J. Cohen (Basingstoke, 2000), pp. 62–79.

21 Daniel Charles, 'The Deluge', in *Lords of the Harvest* (Cambridge, MA, 2001), pp. 236–61.

1 THE ILLUSION OF DIVERSITY: GLOBAL FOOD PRODUCTION AND DISTRIBUTION

1 Mária Ercsey-Ravasz et al., 'Complexity of the International Agro-Food Trade Network and its Impact on Food Safety', *plos one*, vii/5 (2012).

2 See www.who.int/mediacentre, accessed 1 June 2015.

3 Sarah K. Lowder, Jakob Skoet and Saumya Singh, 'What Do We Really Know about the Number and Distribution of Farms and Family Farms Worldwide? Background Paper for The State of Food and Agriculture 2014', *fao esa Working Paper No. 14-02* (Rome, 2014).

4 R. Neumann, 'Chemical Crop Protection Research and Development in Europe', in *Proceedings of the 4th esa Congress*, ed. M. K. van Ittersum and S. C. van de Geijn (Veldhoven-Wageningen, 1996), pp. 49–55.

5 The etc Group, *Putting the Cartel Before the Horse . . . and Farm, Seeds, Soil and Peasants etc: Who Will Control the Agricultural Inputs?* (Ottawa, 2013). Although it is difficult to precisely determine market size and concentration for the overall seed industry, agribusiness consultants Phillips McDougall estimate that the top ten seed companies control three-quarters of the $35 billion global commercial seed market. Also note that although basf's seed sales would not place it in the top ten, it is immersed in seed research and has partnerships with several of the other five Big Six companies as well as investments in several start-up enterprises.

6 In this section I am only referring to the agricultural products of Monsanto, which were retained as part of the new Monsanto that resulted from the original Monsanto/Pharmacia Corporation merger. For a more detailed explanation and timelines of the corporate relationships between Monsanto Company, Pharmacia llc, Pfizer Inc., Solutia Inc. and Eastman Chemical Company, see www.monsanto.com/whoweare, accessed 1 June 2015.

7 Robert T. Fraley et al., 'Expression of Bacterial Genes in Plant Cells', *Proceedings of the National Academy of Sciences*, lxxx (August 1983), pp. 4803–7.

8 Ezra Klein, 'Michael Pollan Thinks Wall Street has Way Too Much Influence over What We Eat', www.vox.com, 23 April 2014.

9 Sallie L. Gaines, 'Monsanto Buying DeKalb Genetics, Delta & Pine', *Chicago Tribune*, 12 May 1998.

10 Quoted in Philip H. Howard, 'Visualizing Consolidation in the Global Seed Industry: 1996–2008', *Sustainability*, I (2009), pp. 1266–87.

11 In 2015, Sygenta ultimately rejected Monsanto's takeover offer. Instead, in February 2016, Sygenta agreed to be acquired by ChemChina in a deal reported to be worth $43 billion. As a state-owned enterprise, ChemChina's takeover underscores the global importance of agribusiness.

12 Rajeev Patel, Robert J. Torres and Peter Rosset, 'Genetic Engineering in Agriculture and Corporate Engineering in Public Debate: Risk, Public Relations, and Public Debate over Genetically Modified Crops', *International Journal of Occupational and Environmental Health*, XIV/4 (2005), pp. 428–36.

13 William Heffernan, 'Report to the National Farmers Union: Consolidation in the Food and Agriculture System', University of Missouri, Columbia, MO, 1999.

14 Nafeez Ahmed, 'UN: Only Small Farmers and Agroecology Can Feed the World', www.permaculturenews.org, 26 September 2014.

15 See www.oxfam.ca/there-enough-food-feed-world, accessed 1 June 2015.

16 Raj Patel, *Stuffed and Starved: The Hidden Battle for the World Food System* (New York, 2012).

17 Howard, 'Visualizing Consolidation', pp. 1266–87.

18 International Service for the Acquisition of Agri-biotech Applications, *ISAAA Brief 46-2013*, www.isaaa.org, 13 February 2014.

19 The Organic and Non-GMO Report, 'Finding Non-GMO Soybean Seed Becoming More Difficult', www.non-gmoreport.com, July/ August 2008.

20 Angelika Hilbeck et al.,'Farmer's Choice of Seeds in Four EU Countries under Different Levels of GM Crop Adoption', *Environmental Sciences Europe*, XXV/1 (2013), p. 12.

21 For India see Glenn Davis Stone, 'Field versus Farm in Warangal: Bt Cotton, Higher Yields, and Larger Questions', *World Development*, XXXIX/3 (2011), pp. 387–98. For South Africa see

Harald Witt, Rajeev Patel and Matthew Schnurr, 'Can the Poor Help GM Crops? Technology, Representation and Cotton in the Makhathini Flats, South Africa', *Review of African Political Economy*, XXXIII/109 (2006), pp. 497–513.

22 Hilbeck et al., 'Farmer's Choice', p. 12.

23 National Research Council, *Impact of Genetically Engineered Crops on Farm Sustainability in the United States* (Washington, DC, 2010).

24 Richard A. Levins and Willard W. Cochrane, 'The Treadmill Revisited', *Land Economics*, LXXII/4 (1996), pp. 550–53.

25 Abraham H. Maslow, *The Psychology of Science: A Reconnaissance* (New York, 1966), p. 15.

26 Thorstein Veblen, 'The Instinct of Workmanship and the Irksomeness of Labor', *American Journal of Sociology*, IV/2 (1898), pp. 187–201.

2 INTELLECTUAL PROPERTY: PROTECTING OR OVERREACHING?

1 'The Universal Declaration of Human Rights', www.un.org, accessed 1 June 2015.

2 John Locke, 'Adam's Monarchy', in *English Prose*, ed. Henry Craik (New York, 1906), p. 185.

3 Audrey R. Chapman, 'A Human Rights Perspective on Intellectual Property, Scientific Progress, and Access to the Benefits of Science', *WIPO/OHCHR: Intellectual Property and Human Rights, A Panel Discussion to Commemorate the 50th Anniversary of the Universal Declaration of Human Rights* (Geneva, 1999), pp. 127–68.

4 Nathanael Johnson, 'Why Vandana Shiva is So Right and Yet So Wrong', www.grist.com, 20 August 2014. This article strikes an interesting middle ground in Johnson's framing of Shiva as being right on the big-picture critique but often wrong about the details, facts and causal mechanisms.

5 Vandana Shiva, *Protect or Plunder?: Understanding Intellectual Property Rights* (London, 2002).

6 Crossing specific parent plants with controlled pollination produces a hybrid seed, often called F1 or F1 hybrids. The seed industry strictly defines these terms and, when they appear in

seed catalogues, they are not applied to plants crossed in the wild.

7 Sally Smith Hughes, 'Making Dollars out of DNA: The First Major Patent in Biotechnology and the Commercialization of Molecular Biology, 1974–1980', *Isis*, XCII (2001), pp. 541–75.

8 See http://supreme.justia.com, accessed 1 June 2015.

9 Jennifer Van Brunt, '*Ex parte Hibberd*: Another Landmark Decision', *Nature Biotechnology*, III (1985), pp. 1059–60. See also the patent application appeal at www.iplawusa.com/resources/ 227_U.S.PQ_443.pdf, accessed 1 June 2015. The 1987 *ex parte Allen* decision went further and ruled that non-naturally occurring, non-human multicellular living organisms, including animals, could be patented.

10 Maryann P. Feldman, Alessandra Colaianni and Connie Kang Liu, 'Lessons from the Commercialization of the Cohen–Boyer Patents: The Stanford University Licensing Program', in *Intellectual Property Management in Health and Agricultural Innovation: A Handbook of Best Practices*, ed. R. T. Krattiger et al. (Oxford, 2007), pp. 1797–807. Available online at www.ipHandbook.org.

11 Hughes, 'Making Dollars out of DNA'.

12 Philip H. Howard, 'Visualizing Consolidation in the Global Seed Industry: 1996–2008', *Sustainability*, I/4 (2009), pp. 1266–87. See updates at http://msu.edu/~howardp/seedindustry.html.

13 CropLife International, 'Cost of Bringing a Biotech Crop to Market', http://croplife.org, accessed 1 June 2015. The average of \$136 million is just a bit more than half of the \$256 million cost of bringing a new conventional chemical crop protection product to market in the period 2005 to 2008 as estimated by a previous study undertaken by Phillips McDougall on behalf of CropLife America and the European Crop Protection Association.

14 Kyle W. Stiegert, Guanming Shi and Jean Paul Chavas, 'Innovation, Integration, and the Biotechnology Revolution in U.S. Seed Markets', *Choices Magazine*, XXV/2 (2010).

15 Before 1 January 1996, the term of a utility or plant patent ended seventeen years from the date of the patent grant. To comply with Article 33 of the TRIPs Agreement resulting from the Uruguay Round Agreements of the General Agreement on

Tariffs and Trade (GATT), patent applications filed on or after 1 January 1996 have a minimum term for patent protection ending no earlier than twenty years from the date the application was filed. Although a patent term begins at filing, enforcement rights only ensue from the date of patent grant, which often takes years.

16 'Roundup Ready Soybean Seed Expiration FAQ', www.soybeans.com/faq.aspx, accessed 1 January 2015.

17 'The Open Source Seed Initiative', http://osseeds.org, accessed 1 June 2015.

18 Jack Kloppenburg, 'The Unexpected Outcome of the Open Source Seed Initiative's Licensing Debate', http://opensource.com, 3 June 2014. Also, see Jack Kloppenburg, 'Re-purposing the Master's Tools: The Open Source Seed Initiative and the Struggle for Seed Sovereignty', *Journal of Peasant Studies*, XLI/6 (2014), pp. 1225–46.

19 See http://osseeds.org/about, accessed 1 June 2015.

20 Melvin S. Nishina et al., 'Production Requirements of the Transgenic Papayas "UH Rainbow" and "UH SunUp"', www.ctahr.hawaii.edu, *New Plants for Hawaii, College of Tropical Agriculture & Human Resources, University of Hawaii at Manoa*, April 1998.

21 The Hawaii Papaya Industry Association that now holds the commercial licensing rights for the Rainbow papaya helped promote the variety started in 1999.

22 Paul Voosen, 'Crop Savior Blazes Biotech Trail, but Few Scientists or Companies Are Willing to Follow', www.nytimes.com, 21 September 2011.

23 Ibid.

24 Amy Harmon, 'A Race to Save the Orange by Altering its DNA', www.nytimes.com, 28 July 2013.

25 Andrew Pollack, 'Gene-altered Apples Get U.S. Approval', www.nytimes.com, 14 February 2015.

26 Amy L. Rice, Keith P. West Jr and Robert E. Black, 'Vitamin A Deficiency', in *Comparative Quantification of Health Risks: Global and Regional Burden of Disease Attribution to Selected Major Risk Factors*, ed. Majid Ezzati et al. (Geneva, 2004), pp. 211–56.

27 Xudong Ye et al., 'Engineering the Provitamin A (Beta Carotene) Biosynthetic Pathway into (Carotenoid-free) Rice Endosperm', *Science*, CCLXXXVII (2000), pp. 303–5.

28 J. Madeleine Nash, 'This Rice Could Save a Million Kids a Year',

www.time.com, 31 July 2000.

29 Vandana Shiva, 'The "Golden Rice" Hoax: When Public Relations Replaces Science', in *Genetically Modified Foods: Debating Biotechnology (Contemporary Issues)*, ed. Michael Ruse and David Castle (Amherst, NY, 2002), pp. 58–62.

30 Michael Pollan, 'The Way We Live Now: The Great Yellow Hype', *New York Times Magazine*, 4 March 2001.

31 R. David Kryder, Stanley P. Kowalski and Anatole F. Krattiger, 'The Intellectual and Technical Property Components of Pro-vitamin A Rice (GoldenRice): A Preliminary Freedom-to-operate Review', *ISAAA Briefs No. 20* (Ithaca, NY, 2000).

32 The Golden Rice Project, http://goldenrice.org.

33 The full text of the ruling is available at *Monsanto Canada Inc.* v. *Schmeiser* (2004) 1 SCR 902, 2004 SCC 34. See http://scc-csc.lexum.com, accessed 1 June 2015.

34 *Organic Seed Growers and Trade Association et al.* v. *Monsanto Company et al.*, Supreme Court Case No. 13-303. See www.supremecourt.gov, accessed 1 June 2015.

35 Adam Liptak, 'Supreme Court Supports Monsanto in Seed-replication Case', *New York Times*, 13 May 2013.

36 General Assembly, 'General Assembly Launches International Year of Quinoa, with Secretary-General Saying Extraordinary Grain Could Have Significant Impact on Anti-hunger Fight', www.un.org, 20 February 2013.

37 Nathanael Johnson, 'A 16th-century Dutchman Can Tell Us Everything We Need to Know about GMO Patents', www.grist.com, 28 October 2013.

38 Paul Thompson, 'The GMO Quandary and What it Means for Social Philosophy', *Social Philosophy Today*, 13 June 2014.

39 Jack Kloppenburg, *First the Seed: The Political Economy of Plant Biotechnology* (Madison, WI, 2005), p. 324.

40 Shiva, *Protect or Plunder?*, p. 18.

41 For a provocative perspective see Michele Boldrin and David K. Levine, 'The Case Against Patents', *Journal of Economic Perspectives*, XVII/1 (2013), pp. 3–22.

42 Jim Wallis and Bill Moyers, *Faith Works: How Faith-based Organizations Are Changing Lives, Neighborhoods and America* (Berkeley, CA, 2001), pp. xvii–xviii.

3 SCARY INFORMATION? LABELLING AND TRACEABILITY

1 See http://justlabelit.org, accessed 1 January 2015.

2 See http://ec.europa.eu, accessed 1 January 2015.

3 Guillaume P. Gruère and S. R. Rao, 'A Review of International Labeling Policies of Genetically Modified Food to Evaluate India's Proposed Rule', *AgBioForum*, x/1 (2007), pp. 51–64.

4 World Health Organization (WHO), *Strategies for Assessing the Safety of Foods Produced by Biotechnology. Report of a Joint FAO/WHO Consultation* (Geneva, 1991).

5 Food and Agriculture Organization of the United Nations, *Biotechnology and Food Safety* (Rome, 1996).

6 Information on the Codex Alimentarius Commission is available at www.codexalimentarius.org; the European Union is also a member organization.

7 Joint FAO/WHO Food Standards Programme, Codex Alimentarius Report of the 36th Session of the Codex Commission on Food Labelling, available at www.codexalimentarius.net.

8 Joint FAO/WHO Food Standards Programme, Codex Alimentarius Report of the 39th Session of the Codex Commission on Food Labelling, available at www.codexalimentarius.net.

9 Jack A. Bobo, 'Two Decades of GE Food Labeling Debate Draw to an End – Will Anybody Notice?', *Idaho Law Review*, XLVIII/2 (2012), pp. 251–65.

10 Ibid.

11 Regulation 1829/2003 of the European Parliament and of the Council of 22 on Genetically Modified Food and Feed, 2003 OJ (L 268) 2 (September 2003), available at http://ec.europa.eu.

12 Ibid., pp. 2–3.

13 Australia New Zealand Food Standards Code 2011, S 1.5.2 (Australia), available at www.comlaw.gov.au.

14 Andrew Szasz, *Shopping Our Way to Safety* (Minneapolis, MN, 2009).

15 Jason Lusk, 'Food Demand Survey (FooDS)', http://agecon.okstate.edu, 16 January 2015.

16 William K. Hallman et al., *Public Perceptions of Genetically Modified Foods: A National Study of American Knowledge and Opinion* (New Brunswick, NJ, 2003).

17 William K. Hallman, Cara L. Cuite and Xenia K. Morin, *Public*

Perceptions of Labeling Genetically Modified Foods Working Paper 2013-01, http://humeco.rutgers.edu, accessed 11 May 2015.

18 Margaret M. Willis and Juliet B. Schor, 'Does Changing a Light Bulb Lead to Changing the World? Political Action and the Conscious Consumer', *Annals of the American Academy of Political and Social Science*, DCXLIV/1 (2012), pp. 160–90.

19 Keith R. Brown, *Buying into Fair Trade: Culture, Morality, and Consumption* (New York, 2003).

20 Julie Guthman, *Agrarian Dreams: The Paradox of Organic Farming in California* (Berkeley, CA, 2004).

21 Nicki Lisa Cole and Keith Brown, 'The Problem with Fair Trade Coffee', *Contexts*, XIII/1 (2014), pp. 50–55.

22 Daniel Jaffee and Philip H. Howard, 'Corporate Cooption of Organic and Fair Trade Standards', *Agriculture and Human Values*, XXVII/4 (2010), pp. 387–99.

23 M. J. Jordán, K. L. Goodner and J. Laencina, 'Deaeration and Pasteurization Effects on the Orange Juice Aromatic Fraction', *LWT – Food Science and Technology*, XXXVI/4 (2003), pp. 391–6.

24 Alissa Hamilton, *Squeezed: What You Don't Know about Orange Juice* (New Haven, CT, 2010).

25 Rachel Schurman and William Munro, 'Targeting Capital: A Cultural Economy Approach to Understanding the Efficacy of Two Anti-genetic Engineering Movements', *American Journal of Sociology*, CXV/1 (2009), pp. 155–202.

26 David Schleifer, 'Categories Count: Trans Fat Labelling as a Technique of Corporate Governance', *Social Studies of Science*, XLIII/1 (2013), pp. 54–77.

4 SCIENTIFIC FALLIBILITY: CONTESTED INTERESTS AND SYMBOLIC BATTLES

1 Paul Sabatier, 'The Method of Direct Hydrogenation by Catalysis', Nobel Lecture, 11 December 1912, available at www.nobelprize.org.

2 Warren Belasco, 'Why Food Matters', *Culture and Agriculture*, XXI/1 (1999), p. 27.

3 William K. Hallman, Cara L. Cuite and Xenia K. Morin, *Public Perceptions of Labeling Genetically Modified Foods Working Paper 2013–01*, http://humeco.rutgers.edu, accessed 1 January 2013.

4 Lee Clarke, *Worst Cases: Terror and Catastrophe in the Popular Imagination* (Chicago, IL, 2006).

5 James M. Acton and Mark Hibbs, *Why Fukushima Was Preventable* (Washington, DC, 2012), available at http://carnegieendowment.org, accessed 15 May 2015.

6 Patrick Sturgis and Nick C. Allum, 'Science in Society: Re-evaluating the Deficit Model of Public Attitudes', *Public Understanding of Science*, XIII/1 (2004), pp. 55–74.

7 Clive James, 'ISAAA Brief 46-2013: Executive Summary', www.isaaa.org, accessed 10 July 2014.

8 Portions of this chapter appear in John T. Lang, 'Genetically Modified Foods: Recent Development', in *Oxford Encyclopedia of Food and Drink in America*, 2nd edn, ed. Andrew F. Smith (New York, 2012), pp. 90–98, and are reproduced by permission of Oxford University Press.

9 Arpad Pusztai, 'Letter From Arpad Pusztai to the Royal Society', 12 December 1999. See www.freenetpages.co.uk.

10 Arpad Pusztai, 'Memorandum', accessed 1 January 2015. See www.freenetpages.co.uk.

11 Stanley W. B. Ewen and Arpad Pusztai, 'Effects of Diets Containing Genetically Modified Potatoes Expressing *Galanthus Nivalis* Lectin on Rat Small Intestine', *The Lancet*, CCCLIV/9187 (1999), pp. 1353–4; Richard Horton, 'Genetically Modified Foods: "Absurd" Concern or Welcome Dialogue?', *The Lancet*, CCCLIV/9187 (1999), pp. 1314–15.

12 Horton, 'Genetically Modified Foods', pp. 1314–15.

13 John E. Losey, Linda S. Rayor and Maureen E. Carter, 'Transgenic Pollen Harms Monarch Larvae', *Nature*, CCCXCIX/6733 (1999), p. 214.

14 Anthony M. Shelton and Richard T. Roush, 'False Reports and the Ears of Men', *Nature Biotechnology*, XVII (1999), p. 832.

15 John M. Pleasants et al., 'Corn Pollen Deposition on Milkweeds in and Near Cornfields', *Proceedings of the National Academy of Sciences of the United States of America*, XCVIII/21 (2001), pp. 11919–24; Karen S. Oberhauser et al., 'Temporal and Spatial Overlap between Monarch Larvae and Corn Pollen', *Proceedings of the National Academy of Sciences of the United States of America*, XCVIII/21 (2001), pp. 11913–18.

16 David Quist and Ignacio H. Chapela, 'Transgenic DNA

Introgressed into Traditional Maize Landraces in Oaxaca, Mexico', *Nature*, CDXIV (2001), pp. 541–3.

17 Matthew Metz and Johannes Futterer, 'Suspect Evidence of Transgenic Contamination', *Nature*, CDXVI (2002), p. 601; Nick Kaplinsky et al., 'Maize Transgene Results in Mexico Are Artefacts', *Nature*, CDXVI (2002), p. 601.

18 David Quist and Ignacio H. Chapela, 'Brief Communications: Reply', *Nature*, CDXVI (2002), p. 602.

19 Scott Smallwood, 'Berkeley Grants Tenure to Critic', *Chronicle of Higher Education,* LI/39 (3 June 2005), p. A8.

20 Lee Clarke, *Mission Improbable: Using Fantasy Documents to Tame Disaster* (Chicago, IL, 1999).

21 William R. Freudenburg, 'Seeding Science, Courting Conclusions: Reexamining the Intersection of Science, Corporate Cash, and the Law', *Sociological Forum*, XX/1 (2005), pp. 3–33; Daniel Lee Kleinman and Jack Kloppenburg Jr, 'Aiming for the Discursive High Ground: Monsanto and the Biotechnology Controversy', *Sociological Forum*, VI/3 (1991), pp. 427–47.

22 Sheila Jasanoff, *Designs on Nature: Science and Democracy in Europe and the United States* (Princeton, NJ, 2005).

23 Thomas Bernauer and Philipp Aerni, 'Competition for Public Trust: Causes and Consequences of Extending the Transatlantic Biotech Conflict to Developing Countries', in *The International Politics of Genetically Modified Food: Diplomacy, Trade and Law*, ed. Robert Falkner (Basingstoke, 2007), pp. 138–54.

5 GETTING BACK ON TRACK: THE TENSION BETWEEN IDEALISM AND DOOM

1 David F. Wallace, 'Consider the Lobster', www.gourmet.com, August 2004.

2 Jack Kaskey, 'Mutant Crops Drive BASF Sales Where Monsanto Denied: Commodities', www.bloomberg.com, 13 November 2013; Jack Kaskey, 'The Scariest Veggies of Them All', www.businessweek.com, 21 November 2013.

3 Tom Philpott, 'Your Vanilla Ice Cream is About to Get Weirder', www.motherjones.com, 4 June 2014; Stephanie Strom, 'Companies Quietly Apply Biofuel Tools to Household Products', www.nytimes.com, 30 May 2014.

4 Diane Farseta, 'The Campaign to Sell Nuclear', *Bulletin of the Atomic Scientists*, LXIV/4 (2008), pp. 38–41.

5 Ibid.

6 Oliver Morton, *Special Report: Nuclear Energy: The Dream that Failed* (London, 2012).

7 Michael Butler, 'Thorium Reactors: Asgard's Fire', www.economist.com, 12 April 2014.

8 Charles Perrow, 'Organizing to Reduce the Vulnerabilities of Complexity', *Journal of Contingencies and Crisis Management*, VII/3 (1999), pp. 150–55.

9 Kristen Purcell, Lee Clarke and Linda Renzulli, 'Menus of Choice: The Social Embeddedness of Decisions', in *Risk in the Modern Age: Social Theory, Science and Environmental Decision-Making*, ed. M. J. Cohen (Basingstoke, 2000), pp. 62–79.

10 Ulrich Beck, *Risk Society: Towards a New Modernity* (London, 1992).

11 Kirk Freudenburg, *The Walking Muse: Horace on the Theory of Satire* (Princeton, NJ, 1993).

12 Dennis H. Wrong, *Power: Its Forms, Bases and Uses* (New Brunswick, NJ, and London, 1995); Steven Lukes, *Power: A Radical View*, 2nd edn (London, 2004).

13 Robert A. Dahl, 'A Critique of the Ruling Elite Model', *American Political Science Review*, LII/2 (1958), pp. 463–9.

14 Molly Ball, 'Want to Know If Your Food is Genetically Modified?', www.theatlantic.com, 14 May 2014.

15 Melanie Warner, 'Wal-Mart Eyes Organic Foods', www.nytimes.com, 12 May 2006.

16 David W. Moore, *The Opinion Makers* (Boston, MA, 2009); George Bishop, *The Illusion of Public Opinion: Fact and Artifact in American Public Opinion Polls* (Lanham, MD, 2005).

17 David B. Lobell et al., 'Prioritizing Climate Change Adaptation Needs for Food Security in 2030', *Science*, CCCXIX/5863 (2008), pp. 607–10.

18 Brian K. Sullivan, Elizabeth Campbell and Rudy Ruitenberg, 'Extreme Weather Wreaking Havoc on Food as Farmers Suffer', www.bloomberg.com, 17 January 2014.

19 Deepak K. Ray et al., 'Yield Trends are Insufficient to Double Global Crop Production by 2050', *PLOS ONE*, VIII/6 (2013), pp. e66428.

20 Patterson Clark, 'Screening Genes for Better Breeding',
    http://apps.washingtonpost.com, 16 April 2014.
21 PR Newswire, 'Cargill Develops Non-GMO Soybean Oil',
    www.digitaljournal.com, 23 June 2014.
22 Jacob Bunge, 'Big Data Comes to the Farm, Sowing Mistrust',
    Wall Street Journal, www.wsj.com, 25 February 2014.
23 Katie Welborn, 'Despite Tech + Ag Hype, Farmers Not as
    Connected as You'd Think', www.civileats.com, 9 January 2015.
24 Michael R. Dimock, 'Bittman Piece Sparks a Clash of Opinion
    and that is a Good Thing!', www.rootsofchange.org, 14 May 2014.
25 Abraham H. Maslow, *The Psychology of Science* (New York, 1966),
    p. 15.
26 Herbert A. Simon, *Administrative Behavior: A Study of
    Decision-making Processes in Administrative Organization*,
    1st edn (New York, 1947); Herbert A. Simon, 'Rational Choice
    and the Structure of the Environment', *Psychological Review*,
    LXIII/2 (1956), pp. 129–38.
27 Rachel Laudan, 'A Plea for Culinary Modernism: Why We
    Should Love New, Fast, Processed Food', *Gastronomica: The
    Journal of Food and Culture*, I/1 (2001), pp. 36–44.
28 Amartya Sen, *Poverty and Famines: An Essay on Entitlement
    and Deprivation* (Oxford, 1981).
29 Nikos Alexandratos and Jelle Bruinsma, *World Agriculture
    toward 2030/2050: The 2012 Revision* (Rome, 2012).
30 David Tilman et al., 'Global Food Demand and the Sustainable
    Intensification of Agriculture', *Proceedings of the National
    Academy of Sciences*, CVIII/50 (2011), pp. 20260–64.
31 Jenny Gustavsson et al., *Global Food Losses and Food Waste:
    Extent, Causes and Prevention* (Rome, 2011).
32 Robert Paarlberg, 'Attention Whole Foods Shoppers',
    www.foreignpolicy.com, 26 April 2010.
33 Ibid.
34 Nathaniel D. Mueller et al., 'Closing Yield Gaps through Nutrient
    and Water Management', *Nature*, CDXC/7419 (2012), pp. 254–7.
35 Rattan Lal, 'Enhancing Crop Yields in the Developing Countries
    through Restoration of the Soil Organic Carbon Pool in
    Agricultural Lands', *Land Degradation and Development*,
    XVII/2 (2006), pp. 197–209.
36 Bernard Malamud, *Dublin's Lives* (New York, 1979).

37 Hunter S. Thompson, *The Rum Diary: The Long Lost Novel* (New York, 1998), p. 5.

BIBLIOGRAPHY

BOOKS AND REPORTS

Alexandratos, Nikos, and Jelle Bruinsma, *World Agriculture toward 2030/2050: The 2012 Revision* (Rome, 2012)

Charles, Daniel, *Lords of the Harvest: Biotech, Big Money, and the Future of Food* (New York, 2002)

ETC Group, *Putting the Cartel before the Horse . . . and Farm, Seeds, Soil and Peasants etc.: Who Will Control the Agricultural Inputs?* (Ottawa, 2013)

Falkner, Robert, ed., *The International Politics of Genetically Modified Food: Diplomacy, Trade and Law* (New York, 2006)

Food and Agriculture Organization of the United Nations, *Biotechnology and Food Safety* (Rome, 1996)

Gaskell, George, and Martin W. Bauer, eds, *Biotechnology, 1996–1999: The Years of Controversy* (London, 2001)

Jasanoff, Sheila, *Designs on Nature: Science and Democracy in Europe and the United States* (Princeton, NJ, 2005)

Kinchy, Abby, *Seeds, Science, and Struggle: The Global Politics of Transgenic Crops* (Cambridge, MA, 2012)

Kloppenburg, Jack, *First the Seed*: *The Political Economy of Plant Biotechnology* (Madison, WI, 2005)

National Research Council, *Impact of Genetically Engineered Crops on Farm Sustainability in the United States* (Washington, DC, 2010)

Paarlberg, Robert, *Food Politics: What Everyone Needs to Know* (New York, 2013)

Patel, Raj, *Stuffed and Starved: The Hidden Battle for the World Food System* (New York, 2012)

Ronald, Pamela C., and Raoul Adamchak, *Tomorrow's Table: Organic Farming, Genetics, and the Future of Food* (New York, 2008)

Schurman, Rachel, and William A. Munro, *Fighting for the Future*

of Food: Activists versus Agribusiness in the Struggle over
    Biotechnology (Minneapolis, MN, 2010)
Shiva, Vandana, Protect or Plunder?: Understanding Intellectual
    Property Rights (London, 2002)
United States Government Accountability Office, Genetically
    Engineered Crops: Agencies are Proposing Changes to Improve
    Oversight, But Could Take Additional Steps to Enhance
    Coordination and Monitoring (Washington, DC, 2008)
World Health Organization, Strategies for Assessing the Safety of
    Foods Produced by Biotechnology. Report of a Joint FAO/WHO
    Consultation (Geneva, 1991)

ARTICLES

Ball, Molly, 'Want to Know If Your Food is Genetically Modified?',
    The Atlantic, www.theatlantic.com, 14 May 2014
Belasco, Warren, 'Why Food Matters', Culture and Agriculture, XXI/1
    (1999), p. 27
Bernauer, Thomas, and Philipp Aerni, 'Competition for Public Trust:
    Causes and Consequences of Extending the Transatlantic
    Biotech Conflict to Developing Countries', in The International
    Politics of Genetically Modified Food: Diplomacy, Trade and Law,
    ed. Robert Falkner (Basingstoke, 2007), pp. 138–54
Bobo, Jack A., 'Two Decades of GE Food Labeling Debate Draw to an
    End – Will Anybody Notice?', Idaho Law Review, XLVIII/2 (2012),
    pp. 251–65
Broad, William J., 'Useful Mutants, Bred with Radiation', New York
    Times, www.nytimes.com, 28 August 2007
Bunge, Jacob, 'Big Data Comes to the Farm, Sowing Mistrust', Wall
    Street Journal, www.wsj.com, 25 February 2014
Chapman, Audrey R., 'A Human Rights Perspective on Intellectual
    Property, Scientific Progress, and Access to the Benefits of
    Science', WIPO/OHCHR, Intellectual Property and Human Rights,
    A Panel Discussion to Commemorate the 50th Anniversary of
    the Universal Declaration of Human Rights (Geneva, 1999),
    pp. 127–68
Charles, Daniel, 'The Deluge', in Lords of the Harvest (Cambridge, MA,
    2001), pp. 236–61
Clark, Patterson, 'Screening Genes for Better Breeding', Washington

Post, http://apps.washingtonpost.com, 16 April 2014

CropLife International, 'Cost of Bringing a Biotech Crop to Market', http://croplife.org, accessed 1 June 2015

Diamond v. Chakrabarty, 447 u.s. 303 (1980), http://supreme.justia. com, accessed 1 June 2015

Ercsey-Ravasz, Mária, et al., 'Complexity of the International Agro-Food Trade Network and its Impact on Food Safety', *plos one*, vii/5 (2012), pp. e37810

European Commission, 'Food Information to Consumer – Legislation', http://ec.europa.eu/food, accessed 1 January 2015

Ewen, Stanley W. B., and Arpad Pusztai, 'Effects of Diets Containing Genetically Modified Potatoes Expressing *Galanthus Nivalis* Lectin on Rat Small Intestine', *The Lancet*, ccliv/9187 (1999), pp. 1353–4

'Ex parte Kenneth A. Hibberd, Paul C. Anderson and Melanie Barker', www.iplawusa.com, accessed 1 June 2015

Falkner, Robert, 'The Political Economy of "Normative Power" Europe: eu Environmental Leadership in International Biotechnology Regulation', *Journal of European Public Policy*, xiv/4 (2007), pp. 507–26

Feldman, Maryann P., Alessandra Colaianni and Connie Kang Liu, 'Lessons from the Commercialization of the Cohen–Boyer Patents: The Stanford University Licensing Program', in *Intellectual Property Management in Health and Agricultural Innovation: A Handbook of Best Practices*, ed. R. T. Krattiger et al. (Oxford, 2007), pp. 1797–807

Freudenburg, William R., 'Seeding Science, Courting Conclusions: Reexamining the Intersection of Science, Corporate Cash, and the Law', *Sociological Forum*, xx/1 (2005), pp. 3–33

Gaskell, George, et al., 'Troubled Waters: The Atlantic Divide on Biotechnology Policy', in *Biotechnology, 1996–2000: The Years of Controversy*, ed. G. Gaskell and M. Bauer (London, 2002), pp. 96–115

Gruère, Guillaume P., and S. R. Rao, 'A Review of International Labeling Policies of Genetically Modified Food to Evaluate India's Proposed Rule', *AgBioForum*, x/1 (2007), pp. 51–64

Hallman, William K., Cara L. Cuite and Xenia K. Morin, *Public Perceptions of Labeling Genetically Modified Foods Working Paper 2013-01*, http://humeco.rutgers.edu, accessed 1 November 2013

Harmon, Amy, 'A Race to Save the Orange by Altering Its DNA',
    New York Times, www.nytimes.com, 28 July 2013

Heffernan, William, 'Report to the National Farmers Union:
    Consolidation in the Food and Agriculture System', University
    of Missouri, Columbia, MO, 1999

Hilbeck, Angelika, et al., 'Farmer's Choice of Seeds in Four EU
    Countries under Different Levels of GM Crop Adoption',
    Environmental Sciences Europe, XXV/1 (2013), p. 12

Horton, Richard, 'Genetically Modified Foods: "Absurd" Concern or
    Welcome Dialogue?', *The Lancet*, CCCLIV/9187 (1999), pp. 1314–15

Howard, Philip H., 'Visualizing Consolidation in the Global Seed
    Industry: 1996–2008', *Sustainability*, I/4 (2009), pp. 1266–87

Hughes, Sally Smith, 'Making Dollars out of DNA: The First Major
    Patent in Biotechnology and the Commercialization of
    Molecular Biology, 1974–1980', *Isis*, XCII (2001), pp. 541–75

International Service for the Acquisition of Agri-biotech
    Applications, *ISAAA Brief 46-2013*, www.isaaa.org, 13 February
    2014

Jaffee, Daniel, and Philip H. Howard, 'Corporate Cooption of
    Organic and Fair Trade Standards', *Agriculture and Human
    Values*, XXVII/4 (2010), pp. 387–99

James, Clive, 'ISAAA Brief 46-2013: Executive Summary', *International
    Service for the Acquisition of Agri-biotech Applications*,
    www.isaaa.org, accessed 10 July 2014

Johnson, Nathanael, 'A 16th-century Dutchman Can Tell Us
    Everything We Need to Know about GMO Patents', *Grist*,
    www.grist.com, 28 October 2013

——, 'Why Vandana Shiva is So Right and Yet So Wrong', *Grist*,
    www.grist.com, 20 August 2014

Joint FAO/WHO Food Standards Programme, Codex Alimentarius
    Report of the 36th Session of the Codex Commission on Food
    Labelling, 2 May 2008, available at www.codexalimentarius.net

——, Codex Alimentarius Report of the 39th Session of the Codex
    Commission on Food Labelling, 13 May 2011, available at
    www.codexalimentarius.net

Kleinman, Daniel Lee, and Jack Kloppenburg Jr, 'Aiming for the
    Discursive High Ground: Monsanto and the Biotechnology
    Controversy', *Sociological Forum*, VI/3 (1991), pp. 427–47

Kloppenburg, Jack, 'Re-purposing the Master's Tools: The Open

Source Seed Initiative and the Struggle for Seed Sovereignty',
Journal of Peasant Studies, XLI/6 (2014), pp. 1225–46

—, 'The Unexpected Outcome of the Open Source Seed Initiative's
Licensing Debate', http://opensource.com, 3 June 2014

Kryder, R. David, Stanley P. Kowalski and Anatole F. Krattiger, 'The
Intellectual and Technical Property Components of Pro-vitamin
A Rice (GoldenRice): A Preliminary Freedom-to-operate
Review', *ISAAA Briefs No. 20* (Ithaca, NY, 2000)

Lang, John T., 'Genetically Modified Foods: Recent Development',
in *Oxford Encyclopedia of Food and Drink in America*, 2nd edn,
ed. Andrew F. Smith (New York, 2012), pp. 90–98

Laudan, Rachel, 'A Plea for Culinary Modernism: Why We Should
Love New, Fast, Processed Food', *Gastronomica: The Journal of
Food and Culture*, I/1 (2001), pp. 36–44

Levins, Richard A., and Willard W. Cochrane, 'The Treadmill
Revisited', *Land Economics*, LXXII/4 (1996), pp. 550–53

Liptak, Adam, 'Supreme Court Supports Monsanto in
Seed-replication Case', *New York Times*, 13 May 2013

Losey, John E., Linda S. Rayor and Maureen E. Carter, 'Transgenic
Pollen Harms Monarch Larvae', *Nature*, CCCXCIX/6733 (1999),
p. 214

Monsanto Canada Inc. v. Schmeiser (2004) 1 SCR 902, 2004 SCC 34.
See http://scc-csc.lexum.com/scc-csc/scc-csc/en/item/2147/
index.do, accessed 1 June 2015

Nash, J. Madeleine, 'This Rice Could Save a Million Kids a Year',
Time, www.time.com, 31 July 2000

*Organic Seed Growers and Trade Association et al. v. Monsanto
Company et al.*, Supreme Court Case No. 13-303. See www.
supremecourt.gov, accessed 1 June 2015

Paarlberg, Robert, 'Attention Whole Foods Shoppers', *Foreign Policy*,
www.foreignpolicy.com, 26 April 2010

Patel, Rajeev, Robert J. Torres and Peter Rosset, 'Genetic Engineering
in Agriculture and Corporate Engineering in Public Debate:
Risk, Public Relations, and Public Debate over Genetically
Modified Crops', *International Journal of Occupational and
Environmental Health*, XIV/4 (2005), pp. 428–36

Pollan, Michael, 'The Way We Live Now: The Great Yellow Hype',
New York Times Magazine, 4 March 2001

Quist, David, and Ignacio H. Chapela, 'Transgenic DNA Introgressed into Traditional Maize Landraces in Oaxaca, Mexico', *Nature*, CDXIV (2001), pp. 541–3

——, 'Brief Communications: Reply', *Nature*, CDXVI (2002), p. 602

Ray, Deepak K., et al., 'Yield Trends are Insufficient to Double Global Crop Production by 2050', *PLOS ONE*, VIII/6 (2013), pp. e66428

Regulation 1829/2003 of the European Parliament and of the Council of 22 on Genetically Modified Food and Feed, OJ (L 268) 2 (September 2003), available at http://ec.europa.eu, 18 October 2003, pp. 2–3

Schurman, Rachel, 'Fighting "Frankenfoods": Industry Opportunity Structures and the Efficacy of the Anti-biotech Movement in Western Europe', *Social Problems*, LI/2 (2004), pp. 243–68

Schurman, Rachel, and William Munro, 'Targeting Capital: A Cultural Economy Approach to Understanding the Efficacy of Two Anti-genetic Engineering Movements', *American Journal of Sociology*, CXV/1 (2009), pp. 155–202

Shiva, Vandana, 'The "Golden Rice" Hoax: When Public Relations Replaces Science', in *Genetically Modified Foods: Debating Biotechnology (Contemporary Issues)*, ed. Michael Ruse and David Castle (Amherst, NY, 2002), pp. 58–62

Stiegert, Kyle W., Guanming Shi and Jean Paul Chavas, 'Innovation, Integration, and the Biotechnology Revolution in U.S. Seed Markets', *Choices Magazine*, XXV/2 (2010)

Stone, Glenn Davis, 'Field versus Farm in Warangal: Bt Cotton, Higher Yields, and Larger Questions', *World Development*, XXXIX/3 (2011), pp. 387–98

Strom, Stephanie, 'Companies Quietly Apply Biofuel Tools to Household Products', *New York Times*, www.nytimes.com, 30 May 2014

Sturgis, Patrick, and Nick C. Allum, 'Science in Society: Re-evaluating the Deficit Model of Public Attitudes', *Public Understanding of Science*, XIII/1 (2004), pp. 55–74

Thompson, Paul, 'The GMO Quandary and What it Means for Social Philosophy', *Social Philosophy Today*, 13 June 2014

Tilman, David, et al., 'Global Food Demand and the Sustainable Intensification of Agriculture', *Proceedings of the National Academy of Sciences*, CVIII/50 (2011), pp. 20260–64

Van Brunt, Jennifer, '*Ex parte Hibberd*: Another Landmark Decision', *Nature Biotechnology*, III (1985), pp. 1059–60

Voosen, Paul, 'Crop Savior Blazes Biotech Trail, but Few Scientists or Companies are Willing to Follow', *New York Times*, www. nytimes.com, 21 September 2011

ORGANIZATIONS

Codex Alimentarius
Codex Secretariat
FAO, Viale delle Terme di Caracalla
00153 Rome, Italy
+39 06 5705.1
www.codexalimentarius.org

Council for Biotechnology Information
1201 Maryland Avenue, SW
Suite 900
Washington, DC 20024, USA
+1 202 962 9200
www.whybiotech.com

Food and Agriculture Organization of the United Nations
Viale delle Terme di Caracalla
00153 Rome, Italy
+39 06 57051
www.fao.org

Greenpeace International
Ottho Heldringstraat 5
1066 AZ Amsterdam
The Netherlands
+31 (0) 20 718 20 00
www.greenpeace.org

International Service for the Acquisition of Agri-biotech Applications
105 Leland Lab
Cornell University
Ithaca, NY 14853, USA
+1 607 255-1724
www.isaaa.org

Non-GMO Project
1155 N State Street, Suite #502
Bellingham, WA 98225, USA
+1 360 255 7704
www.nongmoproject.org

World Health Organization
Avenue Appia 20
1211 Geneva 27
Switzerland
+41 22 791 21 11
www.who.int

ACKNOWLEDGEMENTS

Though writing is a solitary endeavour, I have never been alone while writing this book. And for that, I have to thank all of my friends and family that have encouraged and motivated me. The contributions of others to this book are many and have taken numerous forms, from friendships that will always be cherished to brief but provocative conversations.

Lee 'Chip' Clarke has been a truly giving and inspirational mentor. I certainly owe my deepest intellectual debt to him. But beyond the professional realm, he is a cherished friend. For that, I am truly grateful.

Bill Hallman generously offered to let me learn about GM food when I was a graduate student, working at the Food Policy Institute at Rutgers University. This is where I first learned the thrill and energy of being in a research group. This appreciation for bringing in methods, models and thinking from distinct disciplinary perspectives was expanded even further during my brief stint as a visiting scientist at the Research Centre Jülich-Helmholtz Gemeinschaft in Jülich, Germany, working with Hans Peter Peters, Magdalena Sawicka and their colleagues. By treating me as a fully capable colleague, my academic mentors were models of enthusiasm for intellectual exploration and creativity.

I am thankful to the stimulating environment in the Sociology Department at Occidental College made up of Dolores Trevizo, Jan Lin, Lisa Wade, Richard Mora, Danielle Dirks and Krystale Littlejohn. I would be remiss if I didn't also mention several other people at Occidental who have been invaluable. In particular, I am grateful for conversations and the opportunity to teach with David Kasunic and Carmel Levitan. I've also been fortunate to be surrounded by gifted undergraduate students: Margaret de Larios, Sidney Mathews, Larissa Saco and Elana Muldavin have all provided insightful feedback and research assistance during this work.

I became enthralled with food as a site for sociological exploration when I first attended the joint annual meetings of the Agriculture, Food, and Human Values Society (AFHVS) and the Association for the Study of Food and Society (ASFS) in Austin, Texas, in 2003. I continue to be astounded by the collegiality and generosity of my colleagues in those groups.

I have been fortunate to have the caring support of my friends and family, including my mom, sisters, nieces and nephews, who have wondered when I'm going to finish writing about GM food so I can just go have a nice meal with them. Though I will continue to write, I always look forward to my time with them.

Alyssa Lang, my amazing wife and essential partner, enthusiastically supports all of my new projects. She has accompanied me every step of the way along this journey. Her love and support continues to make a truly wonderful home. Last but not least, my dear son, Beckett, is a source of enormous pride and joy.

INDEX